THE QUEST FOR SPEED

·K·U·P·E·R·A·R·D·

Published in Great Britain by
Kuperard, an imprint of Bravo Ltd
59 Hutton Grove, London N12 8DS
www.kuperard.co.uk
Enquiries: office@kuperard.co.uk

Copyright © 2010 Bravo Ltd

All rights reserved. No part of this publication may be reproduced, stored in a retrieval system, or transmitted in any form or by any means without prior permission in writing from the Publishers.

Series Editor Geoffrey Chesler
Design Bobby Birchall

ISBN 978 1 85733 496 8

British Library Cataloguing in Publication Data
A CIP catalogue entry for this book
is available from the British Library.

Printed in Malaysia

Cover image: Airport tunnel, © *szefei/iStockphoto.com*

Images reproduced under Creative Commons Attribution ShareAlike licence 3.0: page 47 © Lokilech; 48 © Mikani; 50 © Lars-Göran Lindgren Sweden; 57 © Georges Jansoone; 59 © Sven Littkowski/Forum Navis Romana; 128 © Giku; 143 © Greg Heartsfield; 147 © AElfwine; and 158 © Husky (original)

Images reproduced under Creative Commons Attribution ShareAlike licence 2.5: page 73 © AYArktos; and under Creative Commons Attribution ShareAlike licence 2.0: page 44 © Mark McArdle

Image on page 37 courtesy of the State Archives of Florida

Simple Guides »

THE QUEST FOR SPEED

Peter Gosling

⊗ Contents

Introduction	8
1. Speed Over Land	10
▪ On Foot or by Wheel	11
- Early roads	11
- Stagecoaches	12
- Tools of the trade	15
▪ Steam Power – Britain and America on the Move	17
- Railway travel in England	19
- Isambard Kingdom Brunel	23
- The American railroads	24
- Gathering steam	26
▪ Twenty-First-Century Railways	28
- The Channel Tunnel	28
- High-speed trains	29
▪ The Internal Combustion Engine	32
- The horseless carriage	32
- Racing improves the breed	33
Brooklands	36
Indianapolis	38
Le Mans	39
Grand Prix races	40
▪ Two Wheels	46
- Bicycles	46
- Motorbikes	49
- Motorcycle racing	50
2. Speed Over Water	52
▪ From Dug-Out to Sail	52
- Improvements in navigation	55
▪ When Sail was King: the Clippers	57
▪ When Sail was King: the Fighting Ships	59
▪ The First Steamers	61
▪ The Atlantic Crossing	64
- The slave trade	66
- Cutting the journey time	67
- Rowing across	68
▪ The Australia Route	70

- Round the World Sailing Races   71
  - The great sailing races   72
- High Speed on the Water – Power Boats   73
- Ground Effect Vehicles   74

3. Speed Through the Air: the Early Days   76
- Man's Desire to Fly   76
- Balloons   78
- Dirigibles and Airships   81
- Powered Flight   85
  - Aeroplanes   86
  - The Schneider Trophy races   90
- Crossing the Channel by Air   95
- The First Passenger-Carrying Flights: 1918–52   96
- Opening up the World—the Birth of
  - Long-Haul Flights: 1918–49   98
- Flying Eastwards: 1918–39   103
- Trail Blazers and Record Breakers: 1925–37   104
  - Success for Alan Cobham   104
  - 'Amy, Wonderful Amy'   105
  - Alex Henshaw   107
  - Charles Lindbergh   108
  - Amelia Earhart   109
- The Challenge of the Pacific: 1932–39   112
- West to East by Imperial Airways: 1931–39   114

4. Speed Through the Air: the Age of the Jet   118
- Designing the Jet   120
- The Jet Fighters   126
- Breaking the Sound Barrier   129
- The First Commercial Jet Services   134
- Concorde   135
- Air Speed Records   138
- Measuring Speed   139
- Variations on the Theme of Flight   140
  - Rotating-wing machines   140
  - Flying cars   144

5. The Future — 146
- Rockets — 146
    - The Scramjet — 150
- Space Travel — 152
    - Moon landings — 152
    - Visiting Mars — 153
    - The Space Shuttle — 154
    - Commercial space travel — 155
- The Future of Travel: Faster, Further, Smarter? — 157
- Or Will it be Greener? — 159
    - Alternative sources of power — 160

Further Reading — 165
Index — 166

List of Illustrations

Coach-and-four, 1845	13
Nineteenth-century stagecoach	14
Hero of Alexandria's *aeolipile*	17
Diagram of Newcomen's steam engine	18
The Rocket steam locomotive	22
Isambard Kingdom Brunel	23
Driving of the golden spike at Promontory Point	26
Eurostar approaching Chambery	30
The Benz Viktoria	32
Count Louis Zborowski in *Chitty Bang Bang*	37
Malcolm Campbell driving *Blue Bird* in 1935	37
Opel Grand Prix racing car, c. 1913	41
Lewis Hamilton in the Canadian Grand Prix, 2008	44
Donald Campbell's 1964 *Bluebird*	45
Wooden *draisine*, c. 1820	47
Tour de France, 2009	49
The 1914 Triumph TT	50
Viking *drakkar* longship	53
Woodblock print of Zheng He's ships	54
Diagram of a sextant	55

Clipper ship *Thermopylae*	57
Roman galley of 210 BCE	59
The Tudor warship *Mary Rose*	61
The steam turbine-powered *Turbinia*	63
The *Mayflower*	65
The Sirius with steam paddles and sails	67
Model of power boat *Spirit of Australia*	73
The Montgolfier balloon	79
The dirigible Zeppelin III	82
The *Hindenburg* bursting into flames	84
The *Flyer* taking off at Kitty Hawk	86
Louis Blériot	95
Handlley Page long-range biplane	98
Amy Johnson	106
Charles Lindbergh	108
Amelia Earhart	111
Sikorsky S-42 flying boat	113
Short C-Class flying boat	115
Spitfire IIA P7350	118
Gloster E. 28/39	121
Messerschmitt Me 262	123
Whittle jet engine	125
Diagram of operation of a turbofan engine	128
Breaking the sound barrier	129
The Bell X-1 in flight	130
Chuck Yeager next to the Bell X-1	131
Quantas Boeing 747	134
Concorde	136
Leonardo's drawing of a helicopter	141
The V-22 helicopter taking off	143
V-2 Rocket in Peenemünde Museum	147
Messerschmitt Me 163 rocket-propelled fighter	148
NASA experimental scramjet	151
'Buzz' Aldrin walking on the Moon	152
Atlantis breaking the sound barrier	155
Model of Boeing 787 Dreamliner	158
The 'Blue Marble': earth from space	164

⏵ Introduction

One morning in 2008 we drove to our local railway station at Stamford, in Lincolnshire, to begin a journey. We were going to visit our daughter and her family in Voorschoten, a small town in the Netherlands about ten miles from The Hague. We took the train to the airport and flew to Amsterdam, where our daughter met us and drove us to Voorschoten. Travelling time: 5 hrs and 19 mins.

If we had done the same journey one hundred years earlier, in 1908, we would first have travelled by train from a station near our home to Harwich, on the Essex coast, which would have taken five hours. We would then have taken a ferry to the Hook of Holland, near Rotterdam, which would have taken six hours. From there we would have travelled to Voorschoten by steam trams – light railways powered by steam locomotives.

Had we had lived in London at that time, we would have taken one of the twice-daily boat trains to Harwich. It would have been driven straight on to a waiting ship, off again at our port destination, and on to Holland Spoor station in The Hague. We would then have taken a steam tram to Voorschoten. Whatever the route, there would have been much waiting about, and it is possible that the whole trip would have taken up to two days.

Going back another hundred years, to 1808, there would have been no such problems, because the journey would never have taken place: few people would have thought of moving abroad, and we would not have considered visiting them. Our daughter would almost certainly have lived in the same town as we did, in an adjacent street. These snapshots of the same journey show how quickly the speed and technology of travel have moved on, bringing dramatic changes to our lives.

Our quest for speed has been driven by two desires. The first is the wish to go as fast as possible by any means at our disposal, and preferably faster than anybody else. The second, more practical, desire is seen in the improvement in the quality of our transport and travel. The two are closely linked: by pushing technology to the limit, the improvements have filtered down to everyday means of transport. The object of this book is to show how these two ends have been achieved in travel on land, by sea and in the air, since they are all, in many ways, interconnected. We also look at some of the ways that travel will change in the future. As the technology has improved, so times and distances have shrunk, and the world has become a smaller place.

Chapter 1

Speed Over Land

On Foot or by Wheel

The discovery of the wheel was Man's first technological breakthrough. If we look at all the different methods that save us having to walk everywhere – the machines that power them and the machines that themselves make the machines – they all need wheels. The oldest known wheel was found in Mesopotamia, in modern Iraq, and is believed to be over five and a half thousand years old.

In the beginning someone must have realised that a heavy object could be moved more easily if a fallen log, for example, was placed under it and rolled along. Doubtless someone else then had the bright idea of moving heavy objects on a sledge, with a series of rollers underneath, that were expelled at the rear as the sledge moved along. The expelled rollers could then be reinserted at the front of the moving object, and so on. The next improvement would have been the creation of axles, when it was discovered that, as the sledge moved over the rollers, it wore a groove along the inside length of each roller, leaving a 'wheel' standing proud at each end. Someone must have realised that if he could fix an axle beneath the sledge, with a wheel mounted at each end and rotating about it, he had a cart. After that, the next problem that needed to be solved was how to make the cart turn corners.

Early roads
Early men travelled along well-trodden tracks that were used to move sheep and cattle from place to place. As time went on these tracks became easier to navigate, as stones were cleared and trees cut down to make way for travellers. Such paths have been found in all parts of the world – in Europe, Asia and the United States. As wheeled transport of a primitive kind appeared, there came a need for better roads. Five thousand years ago a few simple brick paved roads appeared. Then, some two and a half thousand years ago, as the Romans conquered new territories, they needed to move their armies quickly across occupied lands. Their roads had to be capable of carrying chariots and carts, as well as their foot soldiers. The resulting roads were great feats of engineering, some of which are still visible today in Europe and Asia Minor.

Intended as highways, Roman roads were raised up on a cambered bank of material dug from roadside ditches. In general there were three layers: a layer of large stones, covered by a second layer of smaller stones, with a top layer of gravel or small stones. Each layer was between two and twelve inches deep.

The modern roads we enjoy today are there largely thanks to John McAdam, the inventor of a road construction method that was not only cheap, but also very effective, by providing a surface that made travelling more comfortable. The road construction allowed water to drain off by raising it above the surrounding area with layers of stones and side ditches. The layers could then be compacted by a heavy roller. The technique proved to be so effective that it was copied all over the world.

> John Loudon McAdam (1756–1836), Scottish engineer and road-builder, pioneered the process known as 'macadamising', for building roads with a smooth, hard surface that was more durable and less muddy than soil-based tracks.

Stagecoaches

For thousands of years, if a man needed to travel, he would have to walk. Or, if he had enough money, he would ride his horse. If he was really wealthy, he could travel in a coach pulled by horses. This meant that the majority of people lived in small communities and seldom strayed far from them. However, by the late seventeenth century carriages that had previously been conveyances solely for the rich became more readily available as 'public coaches', for hire. In England, these travelled between London and cities such as Oxford on a daily service; journeys to York, Chester and Exeter would take up to four or five days. Each coach could carry

Coach-and-four, 1845. Only the wealthy could afford their own carriages

up to six people, and travel was priced at a few pence per mile. Coaches travelling up the Great North Road (now known as the A1), between London and York, would stop to change horses at the George Hotel in Stamford, in Lincolnshire. This famous hotel still exists, and possesses rooms called 'The York' and 'The London', where passengers would rest according to the direction of their travel. Travellers from London to Edinburgh would also have stopped here on the ten-day trip to Scotland.

> Improvements in the quality of roads – that made coach travel faster, cheaper and probably less uncomfortable – can be demonstrated by comparing a journey of one hundred miles down the Great North Road, to London, over a century. In 1685 it took two days and cost twenty shillings but, one hundred years later, the same journey took only one day and cost sixteen shillings.

As the quality of the roads improved, thanks to the work of McAdam and others, the journey time of the London mail coach to Holyhead (the Welsh port for crossings to Ireland) was cut, in the 1820s, from 45 hours to just 27 hours, with mail coach speeds rising from an average of 5–6 mph, to 9–10 mph. And, by 1843, the London to Exeter mail coach could complete the 170 miles in 17 hours.

⊙ Nineteenth-century stagecoach

In America, a sophisticated network of stagecoach routes was created in eastern and southern states, but it was in the west that stagecoaches, beloved of cowboy films, were a necessity. The distances that they had to travel were huge and the landscape rugged and dangerous. In 1858, John Butterfield began an overland stage line connecting St. Louis, Missouri, with San Francisco, California, by way of El Paso, Texas. The service ran via Tucson, Arizona, and Los Angeles, California, in those days small and isolated townships. Butterfield's company had won a contract worth $600,000 a year to deliver mail from St. Louis to San Francisco. Within a year it was also providing suitable stops to feed and rest travellers over the 2,795-mile length of the overland mail service. The whole journey took three weeks, weather and Native Americans permitting.

Stagecoach services in America also continued for much longer than in Europe. Even though railways were introduced into the country as early as the 1830s, mail coaches continued to be used for many more years, even into the 1920s, servicing small and isolated communities.

Tools of the trade

The great flowering of technical and economic developments that began in Britain in the late eighteenth century did not happen overnight, but occurred slowly over a number of years. The coming together of science and technology was the result of a long historical process, and the engineers and scientists who enabled the Industrial Revolution to take place needed particular tools to allow them to develop and test their ideas.

The early growth of technology had been hampered by the fact that the Roman numerical system had remained in use until about 900 CE, long after the fall of the Roman Empire. The Hindu–Arabic numeral system, based on ten digits, was first used in Europe around 1000 CE. This made it possible for arithmetic to be performed more easily than with Roman numerals—although financial operations continued to be carried out in the 'old' way until Tudor times.

Our modern decimal place value system was first used by mathematicians in India, probably by the sixth century CE, possibly even earlier, and their system was recognised by Arab and Persian mathematicians in Baghdad, particularly Mohammed ibn-Musa al-Khowarizmi, in the ninth century CE. He wrote two books on algebra and the

Indian numbering system, and the numerals he used have come to be known as Arabic numerals.

Al-Khowarizmi also introduced a form of decimal point to distinguish between whole numbers and fractions, and the word *algorithm* is derived from his name. His algebraic treatise *Hisab al-jabr w'al-muqabala* gives us the word *algebra*, and is believed to be the first book to be written on the subject. The books were subsequently translated into Latin in the twelfth century by Adelard of Bath, enabling European scholars to learn from them.

The scientists of the Islamic world realised that, in order to understand how the world works, they needed to be able to measure the characteristics of everything. This required a means of expressing those characteristics and a method for manipulating them meaningfully.

In England, these concepts were developed further by Isaac Newton (1643–1727), who formulated the mathematical tools later used by engineers. Numbers could now be manipulated according to rules (algorithms), and the characteristics of everything that was measurable could be recorded and put to use. That ability, together with the practical information regarding, for example, the performance of steam under pressure, could be measured, examined and tested. Newton's life's work was concerned with mathematical physics, including defining: the laws of motion; the conservation of momentum; the properties of gases, liquids and solids; and the differential and integral calculus that kick-started the Industrial Revolution.

Steam Power – Britain and America on the Move

It was not the engineers of the Industrial Revolution in the eighteenth century CE who first realised that steam could be a source of power. At some point between the first and the third centuries CE (the dates are disputed), the Greek scientist Hero of Alexandria found that, if steam were made to escape through the end of a small tube, the force of its expulsion resulted in a backward pressure on whatever the tube was fixed to. We now know that that is the principle by which a jet engine works – Hero was ahead of his time. He built a device called an *aeolipile*, which consisted of a metal sphere mounted on a pair of bearings about which it could rotate. The sphere could be partly filled with water, and on each side of the axis of rotation were two thin tubes. The water was heated by an external source and, when it boiled, the steam emerged from the tubes and the sphere started to spin round. History does not recount if the *aeolipile* was ever put to any use, but it worked – although the concept was not rediscovered until almost two thousand years later.

⊙ *Hero's* aeolipile

The dawn of a new age: the Industrial Revolution
By 1800 a new breed of workers was emerging, called 'engineers'. They built bridges, tunnels and railways. The age of steam had arrived.

In mining, the pumping of water out of mines had been a problem that had faced engineers for many years. The earliest uses of pumps using a cylinder and piston were in Greek and Roman times; they were also used in China in the eleventh century.

Tin mining had always been a major industry in Cornwall, but flooding limited the depth at which the mineral could be mined. However, in 1712, Thomas Newcomen perfected a practical steam engine, using a beam, for pumping water. It employed a cylinder containing a moveable piston, connected by a chain to one end of a rocking beam that worked a mechanical lift pump from its opposite end. The top of the power cylinder was open to the atmosphere; steam was introduced at the top stroke to the underside of the piston, and then water was sprayed in, condensing the steam and creating a vacuum. Atmospheric pressure, acting on the upper side of the piston, drove it down.

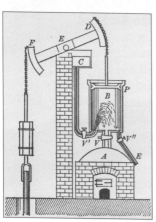

⊙ *Diagram of Newcomen's atmospheric steam engine*

The system worked, despite being rather inefficient, until, in 1776, James Watt produced a better derivative of the Newcomen engine with a separate condenser. In order to reduce the losses in the working of the steam in the steam cylinder, it was necessary to keep the cylinder as hot as the steam that entered it. Watt separated the steam from the cylinder by injecting the cooling water spray into a second cylinder, the condenser, connected to the main one. When the piston had reached the top of the cylinder, a valve was closed, and a second valve, controlling the passage to the condenser, was opened. External atmospheric pressure would then push the piston towards the condenser. Thus the condenser could be kept cold and under less than atmospheric pressure, while the cylinder remained hot.

Railway travel in England

Using tracks to move wagons about began in earnest with the Wollaton Waggonway, Nottinghamshire, in 1603. The tracks were made of wood, with flanged wooden wheels running on an edge rail. In 1768 the first cast iron rails for railways were made in Coalbrookdale, Derbyshire, for use in coalmines. Then, in 1803, the Surrey Iron Railway opened as the first public railway in Britain. The wagons were pulled by horses and ran carrying a few passengers.

Another Cornishman, Richard Trevithick, designed and built a locomotive known as *Puffing Devil* and, on a February day in 1804, Trevithick's locomotive, with its single vertical cylinder, eight-foot flywheel and long piston-rod, managed to haul ten tons of iron, seventy passengers and five wagons from the ironworks at Penydarren in Wales to a local canal, for transportation to a factory. During the nine-mile journey the locomotive reached speeds of almost 5 mph (8 kmph).

> The use of cast iron is one of the great steps forward in the Industrial Revolution. It was first used in China around 500 BCE and, by the fifth century CE, the Chinese were producing wrought iron and steel.

Numerous experiments were made at this time with varying designs of locomotives and, in particular, the suitability of the tracks they ran on. The problem was in the rails: in the early stages they were made of wood, and later of cast iron, that tended to break up as the weight of the locomotives increased. Trevithick, nevertheless, proved that smooth wheels on smooth track could provide sufficient grip to be practical. Only the quality of the iron used, however, stood in the way of a steam-powered railway becoming a commercial proposition. He was able to demonstrate his ideas to the public in 1808, when he built a circular track in London's Euston Square with one of his locomotives (called *Catch-me-who-can*) pulling passengers in a simple truck, for a fare of one shilling, but the public was not impressed.

Interest in railways then lapsed, although better and more reliable locomotives were being built. But in 1820 wrought iron, and later steel, became available for rails and interest revived, largely thanks to the work of Robert Stephenson and

Company, 'Engine Builders and Mill Wrights'. In 1824 the company won an order for locomotives to be used on the first commercial railway line in England, running between Stockton and Darlington. The line opened in September 1825 and the Stephenson locomotive, named *Locomotion No. 1*, hauled thirty eight wagons and six hundred passengers along the twenty-one mile route. Most of the passengers were accommodated in open wagons but a lucky few rode inside the first ever passenger coach, called *Experiment*. The route was used primarily to transport coal, using a steam engine; passengers travelling on the line were horse-drawn. The distance between the rails was 4 ft 8.5 in (1.47 m), which has remained the standard in most parts of the world.

In 1830 the Liverpool to Manchester line was opened. It was preceded by the Rainhill Trials, which were advertised in the *Liverpool Mercury* in May 1829.

TO ENGINEERS AND IRON FOUNDERS

THE DIRECTORS of the LIVERPOOL AND MANCHESTER RAILWAY HEREBY OFFER A Premium of £500 over and above the cost price; for a LOCOMOTIVE ENGINE, which shall be a decided improvement on any hitherto constructed, subject to certain stipulations and conditions, a copy of which may be had at the Railway Office, or will be forwarded as may be directed on appellation of the same, if by letter, post paid.

HENRY BOOTH, Treasurer

Railway Office, Liverpool, April 25, 1829

There were five entries to the trials and Stephenson, with his *Rocket*, was awarded the contract. The Liverpool to Manchester railway was opened with great ceremony on 15 September 1830. In Liverpool, eight trains lined up – one of which carried the Duke of Wellington, victor of the Battle of Waterloo – ready for a grand parade along the route. First class passengers were accommodated in carriages resembling traditional stagecoaches, while second class passengers sat in open trucks with wooden bench seats. The route was lined by cheering crowds, but the opening was unfortunately marred by an accident halfway along, when the MP for Liverpool, William Huskisson, was run over by *Rocket*, driven by George Stephenson, and died of his injuries – he has gone down in history as the first-ever railway casualty.

⊙ The Rocket *steam locomotive*

By 1840 a railway network was being created throughout the United Kingdom, and the Steam Age had begun. The following table illustrates the rapid increase in the size of the railway network.

Year	Miles of track	Number of passengers
1838	500	5,500,000
1848	5,000	30,000,000
1860	10,000	111,000,000

Within a decade the railways had spread north as far as the Scottish border, south to Brighton, to Liverpool in the west, and to Hull in the north-east. By 1854 lines had spread further in all directions, covering most of the country except Wales and parts of Scotland. At last, ordinary people had something they had never had before – affordable mobility. The effect this had on lives everywhere was huge. Villages became towns, towns became cities, and workers could move to where they were needed.

Isambard Kingdom Brunel

England's engineering genius, Isambard Kingdom Brunel (1806–59), was responsible for the design and construction of the Great Western Railway line that joined London's Paddington Station to the city of Bristol, a distance of 118 miles. The line, commenced in 1838 and completed in 1841, was unique in being built with a broad gauge of 7 ft 0¼ in, as opposed to

⊙ *Brunel in 1857*

the more common 4 ft 8½-in gauge. Brunel not only specified the route of the railway, he also designed the viaducts and tunnels along the way. He later extended the line from Bristol to Exeter, with the 194-mile (312 km) journey from London taking 4½ hours, at an average speed of 43 mph (69 kmph).

An important consideration following the creation of the Great Western Railway was the realisation that clocks in London did not tell the same time as those in Bristol. Because Bristol was west of London, there was a difference of eleven minutes between the two places. This caused great problems in the creation of train timetables for journeys across long distances. To begin with, all the stations along the route had two clocks, one showing 'railway time' and the other 'local time'. However, in December 1852, Bristol accepted as standard the time as set by Greenwich Observatory, and this became the standard throughout the country.

The rapid expansion of railway travel was not, of course, confined to the United Kingdom. The countries of Europe, Asia and the Americas all joined in the great railway expansion, and the world began to get smaller. Railway networks became important instruments of change in very large countries, particularly in America, Russia and China.

The American railroads
The early days of the American railroads were very similar to those across the Atlantic in England. The first attempt to transport heavy goods using rails –

called 'gravity roads', or 'tram roads' – was in 1764, to move military equipment at the Niagara Portage, New York. As elsewhere, horses were used to provide the motive power. In 1826 the first passenger-carrying railroad, three miles in length, appeared in Quincy, Massachusetts. Also in 1826, John Stevens of Hoboken, New Jersey, demonstrated a steam-powered train on a circular track, similar to the one demonstrated in England by Trevithick in 1808 with his *Catch-me-who-can*.

The first railroad to move goods and people by steam opened in 1830, operated by the South Carolina Canal and Railroad Company. From then on more and more companies became involved with the new form of transport, but they were mostly small operations covering short distances. In 1830, the Baltimore and Ohio Railroad came into existence with a charter from the Commonwealth of Virginia. By 1850, 9,000 miles (14,484 km) of track had been laid in the USA and the small companies began to amalgamate. The New York Central Railroad Company was created in 1853 as a result of the merger of a dozen smaller companies.

The Civil War (1861–65) halted the growth of the railroads, although those already in existence played an important role in the conflict. After the war ended, construction of the first railroad linking the east and west coasts of the USA began, with support from President Abraham Lincoln. Union Pacific Railroad started construction in the east, and Central Pacific Railroad started from the west. They met at Promontory Point, Utah, on 10 May 1869,

The ceremony of the driving of the golden spike at Promontory Point, Utah

with a golden spike uniting the two tracks. The railroad was built in part to bind the United States together after the American Civil War, and also to encourage the spread of settlers and the creation of new farmlands.

Year	Miles of track
1850	9,000
1860	31,000
1870	53,000
1880	93,000
1890	130,000

Gathering steam

The table opposite, showing train journeys between England and Scotland, illustrates how the speed of rail travel increased over a period of one hundred and fifty years, starting with early steam-powered trains, followed in 1958 by the first diesel-powered trains, the Intercity 125 diesel service in 1976, and the Intercity 225 electric service in 1990.

Date	Distance	Time	Average speed
1865	400 miles	10 hrs 6 min	38.0 mph (61.1 kmph)
1890	400 miles	8 hrs 6 min	47.0 mph (75.6 kmph)
1928	392 miles*	8 hrs 25 min	48.5 mph (78 kmph)
1938	392 miles*	7 hrs 20 min	54.6 mph (88.7 kmph)
1968	392 miles*	5 hrs 50 min	67.2 mph (108.1 kmph)
2008	392 miles*	4 hrs 20 min	92.3 mph (148.5 kmph)

> The diesel engine was invented by Dr Rudolf Diesel, a German mechanical engineer, and first demonstrated in 1898. Power is provided by compressing vaporised fuel inside a cylinder until it ignites and moves a piston. The first diesel engine was fuelled by peanut oil; bio-fuel was then used until the 1920s, when it was replaced by petroleum-derived diesel oil.

The earlier journeys in the table above were from London Euston to Edinburgh Waverley. The starred journeys departed from London Kings Cross. From 1928 the journey was made on the famous *Flying Scotsman* service of the London and North Eastern Railway (LNER). It always departed at 10.00 a.m. every day, and was traditionally sent on its way by the stationmaster of Kings Cross station wearing full uniform.

LNER was proud that the steam engine that pulled the *Flying Scotsman* service became famous in its own right. The engine, an A4 Pacific class locomotive, built in 1929 and designed by Sir Nigel

Gresley, became an icon in the age of steam. Even its number, 4472, became as well known as its name. In 1934 it became the first steam locomotive to reach 100 mph (160.9 kmph). However, its connection with the *Flying Scotsman* service ended in 1939 when war was declared, and it was assigned more mundane duties. In 1955 it was sold into private hands. The locomotive then had a chequered career with various owners, including tours in Australia (where it created a new record for the longest non-stop run by a steam locomotive, 442 miles) and the United States. Now in the National Railway Museum in York, it is currently being restored.

Twenty-First-Century Railways

The Channel Tunnel

The building of a tunnel connecting the United Kingdom to Europe was first considered over two hundred years ago, but nothing happened at the time, for two reasons. Firstly, the technology did not exist that would allow examination of the geology beneath the English Channel, and secondly, there was a fear that Napoleon might use a tunnel as a means to invade England.

By the 1850s, however, serious consideration was being given to the idea. There were, after all, regular train services from London to Dover, and then from Calais to Paris, so it seemed logical to have a fixed, direct, rail route between the two countries. With the spread of the rail network, engineers had by then acquired years of experience building long tunnels – a tunnel under the Thames and another through the Alps had been successfully built. By the 1870s English and French engineers decided that the tunnel was worth trying. They had been able to examine the seabed of the English Channel and had found that it was possible to bore through the chalk there.

> The first tunnel under the Alps was built by the Swiss in 1708. The Gotthard Tunnel, the first rail tunnel under the Alps, was completed in 1881.
>
> The first tunnel under the River Thames, in London, was built in 1825–43 by Marc Isambard Brunel and his son, Isambard Kingdom Brunel, using his newly invented tunnelling shield technology.

In 1881, trial borings were made on both sides of the Channel but, by 1883, the project was abandoned for political reasons. The project remained on hold until 1975 when excavation restarted. Again it was abandoned as the projected cost rocketed. Thirteen years later work began anew. This time it was serious, and the tunnel was eventually completed in 1994. Now there are direct train services from London to Paris and Brussels, and a drive-on drive-off shuttle service enables cars and goods vehicles to be driven from anywhere in the United Kingdom to anywhere on the Continent.

High-speed trains

The great attraction of the Channel tunnel link is the opportunity to travel, by high-speed train, from London to stations in France, and on by further

A British Eurostar train approaching Chambéry in Savoie, France

high-speed trains to major cities in Europe. The Eurostar and Eurotunnel shuttle services have caused a reduction in the number of flights between the UK and many European cities. The table below shows the times to travel to various destinations from England by high-speed train:

Destination	Train time
Paris	2 hrs 15 mins
Brussels	2 hrs 20 mins
Amsterdam	3 hrs 36 mins
Cologne	4 hrs 00 mins
Avignon	6 hrs 30 mins
Marseilles	7 hrs 16 mins
Barcelona	10 hrs 00 mins

Typical international high-speed train journeys are shown in the next table.

Country	Route	Distance	Speed	Time
France	Lorraine-Champagne	167.6 km (104.1 miles)	279.3 kmph (173.6 mph)	1 hr 36 mins
Japan	Okayama-Hiroshima	144.9 km (90.0 miles)	255.7 kmph (158.9 mph)	1 hr 45 mins
Taiwan	Taichung-Zuoying	179.5 km (111.5 miles)	244.7 kmph (152.1 mph)	1 hr 21 mins
Germany	Frankfurt-Bonn	144.0 km (89.5 miles)	233.5 kmph (145.1 mph)	1 hr 36 mins
Belgium	Brussels-Valence	831.7 km (516.8 miles)	244.6 kmph (152.1 mph)	3 hrs 36 mins
Spain	Madrid-Zaragoza	307.2 km (190.9 miles)	227.6 kmph (141.5 mph)	1 hr 21 mins

Speed records all depend on the type of drive used, and records for the most important methods of powering trains are shown in the table below:

Train type	Country	Speed	Year
3rd rail electric	UK	109 mph (175.38 kmph)	1988
Steam locomotive	UK	126 mph (202.73 kmph)	1938
Diesel-powered HST	UK	149 mph (239.74 kmph)	1987
Gas turbine TGV	France	199 mph (320.19 kmph)	1972
TGV electric pantograph	France	357 mph (574.41 kmph)	2007
Maglev	Japan	363 mph (584.07 kmph)	2003

In a Maglev (magnetic levitation) train arrays of magnets, of like polarity in both the vehicle and guideway, repel each other, producing the lifting force. By continuously changing the polarity in alternate magnets, a series of magnetic attractions and repulsions is created that moves the vehicle along the track.

The Internal Combustion Engine

The horseless carriage

In 1885, in Mannheim, Germany, Karl Benz built the first vehicle to be powered by an internal combustion engine. It was a three-wheeler and, despite his best efforts, no one wanted to buy it. However, his wife and sons secretly drove his car sixty-five miles to the town of Pforzheim to display it, where it generated a lot of interest. However, the first motor car to be sold to a member of the public was actually bought in Paris, in 1888. It was a German-built Benz and it was sold to an Emile Roger.

It was only in 1893 that Karl Benz was persuaded to change his design to one with four wheels – the Benz Viktoria. Success followed and companies in Europe and America began manufacturing cars.

The Benz Viktoria. Benz later merged with Daimler to become Daimler-Benz

The first step in the history of personal travel
The internal combustion engine is a device that generates power by a series of controlled explosions of an inflammable gas inside one or more cylinders, thus moving a piston. The fuel vapour is ignited by an electric spark. In 1902, Benz invented the first commercially viable high-voltage spark plug as part of a magneto-based system. (A magneto generates a high voltage current by moving a magnet within a wire coil. That current then causes the spark plug to spark.)

The worldwide automotive industry grew quickly: 4,000 cars were built in the USA in 1900, and over 180,000 in 1910. Freedom to travel at will was becoming available to more and more people. By 1918 there were 5,000,000 cars in the USA. In 2003 the 107 million households in the United States owned an average 1.9 cars each.

The number of cars in the world has increased by 350 percent over the past forty years. Clearly, the car has come to stay, and it has brought with it serious problems. More roads have had to be built and cars are travelling faster. In many ways the car has become the victim of its own success; gridlocks, traffic jams and holdups are all too common, despite efforts to encourage the free flow of traffic. In many cases average speeds, particularly in cities, have come down. The freedom of the road that was promised in the early days has all but vanished.

Racing improves the breed
With the appearance of practical motor cars came the desire to race them and so, in July 1894, the

first motor race was held. It was organised by a newspaper and was run over eighty miles of uneven, dusty and largely unpaved public roads in France, between Paris and Rouen. The race was won by Jules de Dion, driving a car of his own manufacture. It was followed in June 1895 by the Paris–Bordeaux–Paris Rally, a distance of 732 miles (1,178 km), and was the first Grand Prix race in history. It was won by Emile Levassor, driving a Panhard & Levassor, in 48 hrs 47 mins. Also competing was Edouard Michelin, driving a Daimler. Although he came ninth, his claim to fame is that he was driving the first car fitted with pneumatic tyres – which he had to change twenty-two times during the race. The entrants drove cars powered not only by internal combustion engines, but also by steam engines and electric motors.

Before long, races were being held over longer distances. In May 1903 the Paris–Madrid race took place. Over two hundred cars and sixty motorcycles entered, and the event was notable for the large number of accidents that took place along the route. There were at least eight deaths and half the cars crashed. The organisers, because of the danger both to the competitors and to the crowds lining the route, stopped the race at Bordeaux.

In 1907, the famous Peking–Paris race took place. Five cars set out from Peking, and four made it to Paris amid a tumultuous welcome and worldwide fame. The race showed how man and machine could now go anywhere, and demonstrated that borders between countries would become redundant. The participants had to leave Peking without their passports, which had been confiscated by the Chinese authorities who suspected them of being spies (they also had no interest in promoting the motor-car, the Chinese government having invested in the Trans-Siberian railway). There were no roads as we know them today for the first 5,000 miles (8,000 km) of the race, and the competitors had no maps.

The first man to reach Paris, after forty-four days, was an Italian, Prince Scipione Borghese, who had been the favourite from the outset. His specially commissioned vehicle was powered by a seven-litre grand-prix Itala engine, dropped into a truck chassis made with the lightest, simplest bodywork. The second driver home was a Dutchman, Charles Goddard, driving a Dutch Spyker, who managed to drive single-handedly for a period of twenty four hours, a feat not repeated until the coming of the Le Mans race. Goddard, incidentally, had never driven a car before he set out to take part in the race.

The other competitors were: Auguste Pons, driving a three-wheeler Contal Cyclecar; Georges Cormier driving a De Dion; and Victor Collignon, also in a De Dion. The Cyclecar, a basketwork body on three wheels, crashed en route.

It soon became clear that if car racing was going to become a great spectator sport, races should be held on specially built circuits, and not over public roads, to allow the cars to demonstrate their qualities to as many people as possible.

Brooklands

The Brooklands racing circuit in south London was opened in June 1907. It took the form of a 3¼-mile long banked oval, with two straights. The track was one hundred feet wide and the banking was thirty feet high. On the occasion, three Napier cars were driven around the circuit for twenty-four hours. S. F. Edge drove one of the Napiers for the whole time, covering 1,581 miles at an average speed of almost 66 mph, establishing a record that stood for seventeen years.

The first land speed record to be recorded at Brooklands was in November 1909, when the French driver Victor Hémery drove a 200 hp Benz at 115.93 mph. It was not long before more records were being taken at Brooklands including, in February 1913, Percy E. Lambert, the first person to cover a hundred miles in an hour on the circuit, driving a 4.5 litre Talbot and covering 103.84 miles in the hour. The first two-way land speed record was made in June 1914 by the British driver L. G. Hornstead, driving another 200 hp Benz at a speed of 124.09 mph. ('Two-way' speed is the average speed over two runs in opposite directions, to allow for wind.)

One of the most famous racing cars of the time that raced at Brooklands was *Chitty Bang Bang*, owned and driven by Count Louis Zborowski. It was a monster of a car, powered by a 23 litre six-cylinder Maybach aero engine, and won its first race in 1921. It lapped at 111.92 mph and reached a speed of almost 120 mph on the straight.

Count Louis Zborowski in Chitty Bang Bang *at Brooklands, c. 1921*

Records continued to be made and broken at Brooklands in the years after the First World War. The last land speed record achieved on that track was in May 1922 when Kenelm Lee Guinness, a member of the famous brewing family, drove a 350 hp Sunbeam at a two-way average speed of 135.75 mph. The car, powered by a V12 Sunbeam Manitou aero engine, was soon after acquired by Malcolm Campbell and became his first *Blue Bird* land speed record breaker.

The first British Grand Prix race was held at Brooklands in 1926.

Malcolm Campbell's record-breaking Blue Bird *at Daytona Beach in 1935*

In 1930, *The Daily Herald* newspaper put up a trophy for the fastest driver round the Brooklands track. Kaye Don, the first winner, battled with Tim Birkin to achieve 137.58 mph in his Sunbeam Tiger. In 1932, Tim Birkin took the record to 137.96 mph in his red blower Bentley, and John Cobb finally took the record at 143.44 mph in his Napier Railton. Regarded as the ultimate Brooklands Racing Car, it was powered by a 24-litre Napier Lion engine, and the car's outer circuit record remained unbeaten when racing and record breaking finished at Brooklands in 1939.

Indianapolis

The famous purpose-built racing circuit in America, the Indianapolis Motor Speedway, was built in 1909. Although originally a gravel-covered track, it was quickly paved with 3.2 million bricks – hence its nickname of 'the brickyard'. The track, an oval with two straights and two corners banked at 9°, covers a distance of 2½ miles.

The first five-hundred-mile race held there, two hundred laps of the circuit, was held on Memorial Day, 30 May 1911. This race has been run annually on that day, except during the war years of 1917–18 and 1942–45. First place in the first 'Indy 500' was taken by Ray Harroun, at an average speed of 74.602 mph. He drove the only single-seat car in the race, a Marmon Wasp – all the other thirty-nine entrants drove with a mechanic on board.

Non-American competitors began to make their mark at Indianapolis when Australian Jack Brabham drove a slightly modified F1 Cooper in the 1961 race.

In 1963, British designer and driver Colin Chapman brought his Team Lotus to Indianapolis for the first time, attracted by the large monetary prizes. Racing a mid-engined Lotus, Scotsman Jim Clark was second in his first attempt in 1963, and won the race in 1965, a victory that also interrupted the success of the Australian. The following year saw another British win, this time Graham Hill in a Lola-Cosworth.

The two current records for the Indy 500 are held by the Dutch driver Arei Luyendyk. In May 1996 he made the fastest lap in practice at a speed of 239.26 mph, and went on to win the race at an average speed of 185.981 mph.

Le Mans

Another event with a long and exciting history is the race known as the '24 Heures du Mans'. It has been held at the circuit at the town of Le Mans, in France, since 1923, annually in June. It has been cancelled only twice; once in 1936, and also during the war years of the Second World War.

The Le Mans circuit consists of sections of permanent track and public roads that are temporarily closed for the race. It is a sports car endurance race and is 8.48 miles in length. Traditionally the event began with what became known as the *Le Mans start*, in which cars were lined up alongside the pit wall in the order in which they qualified. The drivers would stand on the opposite side of the track. When the French flag dropped to signify the start, they would run across to their cars and start without assistance before driving away.

This was eventually abandoned for safety reasons – on one occasion a driver got into trouble for, as he jumped into his car, the gear lever disappeared up his trouser leg... Now, a 'rolling start' takes place.

In the early days the race was dominated by cars entered by Bugatti, Bentley and Alfa Romeo. It was the rivalry between the Bugatti and Bentley teams, in particular, that produced great racing. Ettore Bugatti, an Italian working in France, created cars that were works of art, incredibly light and fast, whereas W. O. Bentley designed cars that were rooted in his early days in the railway locomotive industry. Ettore commented, 'Mr Bentley makes very fast lorries!', but the Bentley team won the race five times in twelve years and, in 1929, took the first four places. At that race the winner covered 1,767 miles (2,843 km) at an average speed of 73.6 mph (118.5 kmph).

How average speeds increased

Year	Car	Distance travelled	Average speed
1923	Chanard & Walker	1,367 miles (2,200 km)	57.0 mph (1.71 kmph)
1929	Bentley	1,767 miles (2,844 km)	73.6 mph (118.42 kmph)
1935	Lagonda	1,867 miles (3,004 km)	77.8 mph (125.18 kmph)
1949	Ferrari	1,875 miles (3,017 km)	82.3 mph (132.42 kmph)
1985	Porsche	3,160 miles (5,085 km)	131.7 mph (211.90 kmph)
2008	Audi (Diesel)	3,231 miles (5,200 km)	134.62 mph (216.60 kmph)

Grand Prix races

The first Grand Prix race took place in 1906, on the Le Mans circuit. Further races were held over public roads in France under the auspices of the Automobile Club de France.

Year	Location	Lap length	No. of laps	Fastest lap	Driver & car
1906	Le Mans	64.11 miles (103.15 km)	12	73.7 mph (118.58 kmph)	Ferenc Szisz Renault
1907	Dieppe	47.84 miles (76.97 km)	10	75.4 mph (121.32 kmph)	Felice Nazzaro Fiat
1908	Dieppe	47.84 miles (76.97 km)	10	78.65 mph (126.55 kmph)	Christian Lautenschlager Mercedes
1912	Dieppe	47.84 miles (76.97 km)	12	78.55 mph (126.39 kmph)	Georges Boillot/Peugeot
1913	Amiens	19.65 miles (31.62 km)	29	76.71 mph (123.43 kmph)	Georges Boillot/Peugeot
1914	Lyons	23.38 miles (37.62 km)	20	69.78 mph (112.28 kmph)	Christian Lautenschlager Mercedes

By the 1920s motor racing was becoming more and more popular and in 1926, for example, there were no fewer than thirty-five 'Grand Prix' races in Europe. A variety of vehicle types took part in these races: 'Formula Libre', 'Voiturette' and 'Cyclecar'. Formula Libre class cars had no restriction on

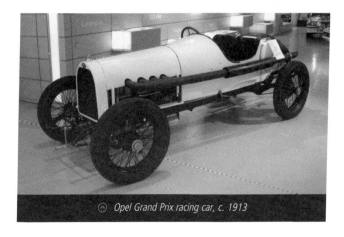

Opel Grand Prix racing car, c. 1913

engine size. Voiturettes were originally cyclecars (mainly three-wheelers), although the term was later used to define cars with engine sizes of no more than 1,500 cc. Typical racing cars in this group were the British ERA and the French Bugatti Type 35. Cyclecars were generally cars with a small engine capacity, between 750 cc and 1,100 cc.

Details of a sample of races run in 1926 are shown below:

Race	Location	Distance	Winner	Car	Fastest lap
Circuito del Pozzo	Verona	156.4 miles (251.65 km)	Alessandro Consonno	Bugatti 35	77.56 mph (124.80 kmph)
Premio Reale di Roma	Valle Guilia	186.4 miles (300 km)	Aymo Maggi	Bugatti 35	65.08 mph (104.71 kmph)
Grand Prix d'Europe	Lasarte	496.34 miles (800 km)	Jules Goux	Bugatti 39A	74.43 mph (119.76 kmph)
Gran Premio de Espana	Lasarte	441.19 miles (709.87 km)	Meo Constantini	Bugatti 35	78.84 mph (126.85 kmph)
Grand Prix de la Marne	Reims	198.84 miles (320 km)	François Lescot	Bugatti 35	70.08 mph (112.76 kmph)
RAC Grand Prix	Brooklands	287.77 miles (463 km)	Senechal/ Wagner	Delage 155B	71.67 mph (115.32 kmph)
Großer Preis von Deutschland	Avus	243.26 miles (391.40 km)	Rudolf Caracciola	Mercedes	83.76 mph (134.77 kmph)
Junior Car Club '200'	Brooklands	201.97 miles (325 km)	Henry Seagrave	Talbot 700	75.56 mph (121.58 kmph)

It is an accepted fact that motor racing is dangerous and, in the early races, deaths on racing circuits, which usually took place over public roads, were considerable: casualties included not only drivers, but also members of the public who strayed too close to the vehicles. The cars themselves were, in those days, difficult and dangerous to drive.

There was no protection for the driver, the wheels were probably shod with solid tyres, and brakes were only fitted on the rear wheels.

For many years, until the 1970s, there was little protection given to the driver, even of Grand Prix cars. Very often the driver sat high in the cockpit and wore a simple crash helmet, with no protective clothing and no seat belt. The result was that many of the drivers who crashed had little chance of escape and were often burnt to death. Two accidents in particular caused much upset, and marked a move to improve the safety of drivers in races. In 1967 Lorenzo Bandini, driving a Ferrari in the Monaco Grand Prix, left the track and his car flipped over. Although Bandini was thrown out; the car caught fire and trapped him beneath it. He died three days later. The following year the brilliant British driver, Jim Clark, was killed driving a Lotus 48 at the Hockenheimring in Germany when he hit a tree. He was trapped and later died on his way to hospital.

Subsequent safety measures included: gravel run-offs adjacent to tracks which slowed cars down that had left the track; where there had been only rope barriers between the cars and crowds in the countryside, steel barriers were erected. The cars themselves became safer, with the driver sitting inside what is known as the reinforced 'tub', tightly belted, wearing flameproof overalls and a helmet, and with restraints against whiplash injuries.

The last fatality on a Grand Prix racing circuit was in 1994 when Ayrton Senna, a triple World Champion, was killed driving a Williams in the San

Lewis Hamilton rounding a bend in the Canadian Gtand Prix, 2008

Marino Grand Prix. Since then there have been many spectacular accidents resulting in injuries from which the drivers completely recovered. Safety is now one of the main considerations in motor racing.

The following chart shows the increase in the land speed record for cars (some steam, some electric, most powered by an internal combustion engine), from 1898 to 1947:

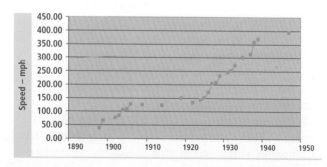

Until 1947 all speed record-holding cars were 'proper' cars, with an internal combustion engine driving the wheels. With the advent of jet propulsion cars became, in effect, 'trollies' that were pushed along by the thrust of a turbojet – or, in one case, a rocket engine.

On 17 July 1964, at Lake Eyre, Australia, Donald Campbell set a record of 403.10 mph (648.72 kmph) for a jet-powered four-wheeled vehicle. Although Craig Breedlove had taken his car *Spirit of America* to 408.31 mph in 1963, his record had originally been considered invalid as the car only had three wheels.

Campbell's 1964 Bluebird with a tail fin and strengthened cockpit canopy

By 1970 the land speed record had been taken over 600 mph (965.6 kmph). Finally, in 1997, RAF pilot Andy Green drove *Thrust SSC* to the first supersonic land speed record of over 760 mph (1223.1 kmph).

At the time of writing there is a project to propel a car to 1,000 mph (1609.34 kmph) with a mixture of a rocket engine, a turbojet and a V-12 racing car engine. The *Bloodhound SSC* is being built in the UK and will attempt the record in 2011.

There is also a land speed record for a wind-powered vehicle, a land yacht. Set on a dry lake in Nevada in 2009 by Britain's Richard Jenkins, it stands at 126 mph (203 kmph). The land yacht has a 26 ft high sail, weighs only 770 lb and produces power equivalent to 435 hp.

> A new attempt is being made on the land speed record for a steam-powered vehicle. The present record stands at 127.7 mph, and was set in 1906! The new car, to be driven by the grandson of Sir Malcolm Campbell, expects to reach 170 mph.

Two Wheels

Bicycles

As roads became easier to travel over, more and more people began to use them for personal as well as commercial transport. Travelling was becoming more of a pleasure than an uncomfortable chore. Personal transport was still confined to horseback but, in 1818, the *hobbyhorse* made its appearance. This was the direct ancestor of the bicycle. Made of wood, with two unsprung wheels and a leather seat, it was propelled in a simple manner by foot power and must have been very uncomfortable.

A similar machine was built at the same time by a German, Baron Karl von Drais, who called it a *dandy horse*. It was similar to the hobbyhorse but

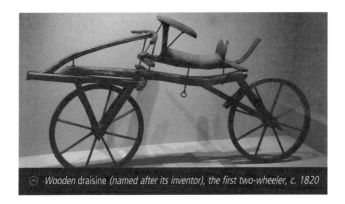

Wooden draisine (named after its inventor), the first two-wheeler, c. 1820

with a steerable front wheel. Although hardly a serious method of transport, it was the beginning of a popular and pleasurable leisure occupation.

Around 1860 someone had the idea of adding a pair of cranks to the front wheel of a dandy horse, making it possible for the rider to propel himself along. The age of the bicycle was dawning.

Since that time the design of the bicycle has changed significantly, with the penny-farthing, safety bicycle, racing bicycle, folding bicycle and the mountain bike all making their appearance. These have all benefited from improvements in technology in the gear systems, wheels, steering, frames and weight of the machines.

Cycle racing takes various forms but the best known, and most prestigious, event is the annual Tour de France, which first ran in 1903. The route taken each year varies, but the 2,200-mile (3,500 km) event, spread over twenty-three stages, always finishes in Paris. It is the rider with the least elapsed time each

Stage Three of the Tour de France, 2009

day who gets to wear the famous *maillot jaune* (yellow jersey), and the times taken to finish each stage, when totalled, determine the overall winner.

Speed records on bicycles are varied, as there are a number of different ways of gaining high speeds. An ordinary domestic bicycle can reach 30 mph (50 kmph) over a short distance on the road, and track riders can touch 40 mph (65 kmph). But the really high speeds are reached in several ways: for example, being paced by a vehicle driving in front of the cyclist; doing a flying lap in a specially prepared streamlined and enclosed machine; or even down a 45° slope aided by gravity. Examples of these (all pedalled bicycles) are:

Date	Rider	Technique	Speed
1923	Charles 'Mile a Minute' Murphy	Paced by a railway train	60 mph (96.5 kmph)
1929	Sam Whittingham	Flying 200m streamlined enclosed	81 mph (130.3 kmph)
1935	Sam Whittingham	Recumbent bike	81 mph (130.3 kmph)
1949	Markus Stoeki	45° snow slope	130.7 mph (210.3 kmph)
1985	John Howard	Paced by 500 hp dragster on Bonneville Salt Flats	152.2 mph (244.9 kmph)

Motorbikes

As the design of bicycles improved, and the technology behind motive power got better, the two were married together in 1868.

The first motorbike was powered not by a petrol engine, but by a steam engine. It was built in 1867 by an American, Sylvester Howard Roper, of Roxbury, Massachusetts, with a wood and iron frame and iron shod wooden wheels. Although the machine was widely demonstrated in the US, it did not catch on, but it did anticipate many modern motorbike features, including the twisting-handgrip throttle control. The Roper bike was powered by a charcoal-fired two-cylinder engine, whose connecting rods directly drove a crank on the rear wheel.

It was a German, Gottlieb Daimler, who designed the first-ever internal combustion-engined motorcycle, in 1895. As well as its main two wheels, it also had a smaller spring-loaded outrigger wheel attached to each side. It was made mostly of wood, with the wheels iron-banded and wooden-spoked, and was powered by a single-cylinder engine.

In 1883 Siegfried Bettmann started a company selling bicycles made by Andrews of Birmingham but, in 1886, he named his company the Triumph Cycle Company and began selling bicycles of his own manufacture. In 1902 he fitted a single-cylinder engine to one of them, and in 1905 marketed the first Triumph motorcycle with a 300 cc, 3 hp engine. And so the motorcycle was born, with more and more companies designing their own versions of the powered bicycle.

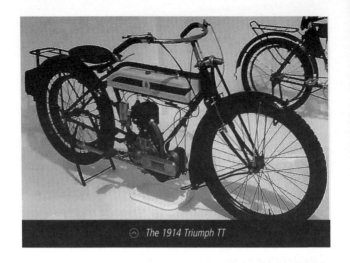
The 1914 Triumph TT

At the outbreak of the First World War, Triumph produced the Model H Roadster with a 550 cc side valve engine and three-speed gears. It became known as the 'Trusty' Triumph, and is recognised as the first modern motorcycle. During the war the company supplied 30,000 machines to the Allied forces.

Motorcycle racing
Motorcycle racing exists in various forms, according to the size of the engine. The famous Isle of Man TT (Tourist Trophy) races give a good indication of the rise in racing speeds on two wheels.

The British government would not allow racing over public roads, unlike many Continental governments, but the self-governing Isle of Man allowed the TT road races as early as 1907, when the original short course round the island was used. The circuit (now known as the Mountain Circuit) was lengthened from 15.8 miles (25.5 km) to 37.73 miles (61 km) in 1911.

The fastest lap in the twin-cylinder class was 42.9 mph (69.0 kmph) and, in the single-cylinder class, was 41.8 mph (67.3 kmph). In 1910 the last race on this course was won at an average speed of 50.63 mph (81.46 kmph). Speeds rose dramatically once the Mountain Circuit came into use, and by 1937 the lap record had reached 90.27 mph. It must be remembered that in the early days, the roads were in poor condition and the bikes were in no way streamlined, with thin section cross-ply tyres and drum brakes.

After the Second World War the TT became one of the World Championship races and, in 1956, a lap speed of 99.97 mph (160.85 kmph) was reached. Eventually, the magic 'ton', one hundred miles per hour, was reached in 1967, when Mike Hailwood set a 108.77 mph (175.04 kmph) record that stood for eight years until it was beaten by Mick Grant at 109.82 mph (176.73 kmph). By now the bikes were fitted with disc brakes, fairings and radial-ply tyres that allowed cornering at speed and at angles of over 70° to the vertical. In 1980 a lap speed of 115.22 mph (185.39 kmph) was reached. Lap records continued to rise until, in 2007, the record jumped again and was taken by John McGuinness in the Senior TT with a time 130.344 mph (209.763 kmph).

⏵ Chapter 2

Speed Over Water

From Dug-Out to Sail

Early civilisations generally developed close to rivers, so that they had access to water for themselves and to what must have been the first 'roads'. There were few land tracks and water transport was far easier and faster than dragging carts overland.

The earliest boats were probably dugouts made from hollowed-out tree trunks, and were certainly used in the Stone Age – the remains of prehistoric dugouts have been found in Germany. Other craft of the time were simple canoes that consisted of a light frame covered in bark or hide, such as the hide-covered canoes used by the Native Americans.

For five thousand years the Ancient Egyptians used ships for trading purposes, on both the Mediterranean and the Red Sea. Records from the fourth century BCE refer to the building of ships 100 cubits (167 ft, 50.9 m) long to trade with the East in spices, sweet-scented woods and fragrant gums. These would have been boats propelled by oars, with possibly a steering oar at the stern. Although much of the traffic was along the River Nile, sea-going ships, which would have had two or three banks of oars and sails made of matting, would have been used on rare occasions.

Boats also formed part of Ancient Egypt's religion. The earliest depictions of the Sun god Ra show him travelling on a reed boat. Even after these had been replaced with wooden boats, 'papyriform' boats made of wood preserved the connection with royalty and gods. These were used in religious events and as funerary or burial boats.

Ships were also built in other countries bordering the Mediterranean and, by 1200 BCE, the Phoenicians and Greeks had begun to construct sailing ships 100 ft (30.5 m) long and capable of carrying 90–180 tonnes of cargo. The Romans built even bigger ships that could carry up to 1,000 people and 1,000 tonnes of cargo.

The eighth century CE saw the rise of the Vikings, who were famous for their longships, the flat bottoms of which allowed them to sail up shallow rivers, and which could even be dragged overland, during their raids across Europe. They were clinker built, with hulls made of overlapping

⊙ *A troop-carrying Viking longship, or* drakkar

planks hand cut with adzes. Called *knorr*, these ships were long and slender and had a large square sail. They were steered with a single oar from the side and were swift and capable of long voyages but they could not carry large cargoes.

By 1200, ships based on the *knorr* were used all over northern Europe. Gradually two castles were added to the bow and stern to provide platforms for soldiers to attack their enemies. A small castle was sometimes added to the mast for bowmen to stand and fire from. Thus they became shorter and more heavily bodied. Until about the middle of the eighteenth century the design of cargo ships and ships of war tended to be similar, as in many cases cargoes had to be defended from attack by pirates.

In 1405 the Chinese Admiral Zheng He assembled a treasure fleet consisting of sixty-two ships; four were huge wooden boats about 400 ft (121.9 m) long and 160 ft (48.77 m) wide. They were the flagships of the fleet assembled at Nanjing along the Yangtze (Chang) River. Also in the fleet were 339-ft (103.3 m) long horse ships carrying only horses, water ships that carried fresh water for the crews, troop transports, supply ships, and warships for offensive

⊙ *Woodblock print thought to represent Zheng He's ships. 17th century*

and defensive needs. The ships carried thousands of tons of Chinese goods to trade with countries visited during the voyage. In the autumn of 1405 the fleet was ready to embark with 27,800 men. Its destination was the south-western coast of India. During the voyage the fleet stopped in Vietnam, Java, and Malacca, and then headed west across the Indian Ocean to Sri Lanka, and then on to Calicut and Cochin. They remained in India to barter and trade from late 1406 to the spring of 1407, when they made use of the monsoon to sail homewards, arriving in 1407.

Improvements in navigation

As journeys by sea were becoming more frequent, and taking place over increasingly large distances, it became essential for a ship's position on the globe to be known accurately by determining its latitude (north–south position), and its longitude (east–west position). Latitude could readily be established by examining the inclination of the sun using a sextant, or its earlier sixteenth century version, the backstaff. But longitude was more difficult to establish. In 1714 the British Parliament passed the

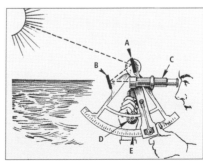

⊙ *Diagram of a sextant*

Longitude Act and a prize of £20,000 was offered for a method of determining longitude to an accuracy of half a degree (or 60 nautical miles) of a great circle.

> **Longitude**
> A line of longitude is a north–south meridian and half of a great circle. Longitude shows your location in an east–west direction, relative to the Greenwich meridian.

The problem was tackled by John Harrison, a joiner and clock maker, who constructed his first chronometer between 1730 and 1735. It was a portable version of his precision wooden clocks, and was spring-driven, but it only ran for one day (the wooden clocks ran for eight days). The moving parts were controlled and counterbalanced by springs. It had no pendulum and so was independent of the pull of gravity. The method of establishing longitude is based on the fact that, for every 15° one travels eastward, the local time moves one hour ahead. Similarly, travelling west, the local time moves back one hour for every 15° of longitude. Therefore, if we know the local times at two points on Earth, we can use the difference between them to calculate how far apart those places are in longitude, east or west.

> **John Harrison**
> The inventor of a device that could accurately measure longitude was born in 1693. He lived in Lincolnshire, and built a series of precision pendulum clocks that were made almost entirely from wood.

By having what were now sophisticated methods for finding location, the opportunities for exploration and trading over long distances were now increased.

When Sail was King: the Clippers

In the late 1840s American shipbuilders responded to the need for a new type of fast cargo ship. The rush to the gold fields of California meant that the miners there needed supplies quickly. Food and drink was shipped in from the east coast of America, and premium prices were paid for delivery by the fastest ships available.

The new breed of ships, known as 'clippers', had to be able to transport their small but valuable cargoes at high speeds over long distances. But it was not only for the voyages south from New York,

The clipper ship Thermopylae, *built in 1868 for the China tea trade*

round Cape Horn, and then north to San Francisco that clippers were used. The routes from England to Australia were becoming more and more important. The China, or Tea, Clippers were used on the trade routes between Europe and the East Indies. One of the most famous of these is *Cutty Sark*, now preserved in dry dock at Greenwich, in the United Kingdom, and currently being refurbished after a fire in May 2007.

> The clippers were long and slim, with a sharp bow for cutting through the water and three tall masts, the height of a twenty-storey building, carrying a large sail area. These sails, together with the racing design of the ship, enabled the ships to 'clip' along faster than any other ship.

Up to the 1840s merchant ships were usually capable of making 5 knots (6 mph/9.6 kmph). Clippers were designed to reach at least 9 knots (10 mph/16.1 kmph). With a trade wind behind them, a speed of 20 knots (23 mph/37 kmph) was possible. In 1854 the clipper *Sovereign of the Seas* recorded 22 knots (25 mph/40 kmph), the fastest speed of any sailing vessel of the time on a trip eastwards to Australia.

> 1 knot = 1.15 mph (1.852 kmph) per hour

The Boston-built *Sovereign of the Seas* created several records early in her life. She was the first ship to travel more than 400 miles (644 km) in twenty-four hours. She also travelled from Honolulu to New York in 82 days and sailed from New York to Liverpool in 13 days 13½ hours.

Many other clippers were capable of high speeds. The *Rainbow*, for example, had a top speed of 14 knots (16 mph/25 kmph) and a large number could reach 18 knots (21 mph/33 kmph) or more. The *Champion of the Seas* made a fast 465-mile run in one day, and there are other cases of clippers sailing over 400 nautical miles in 24 hours.

When Sail was King: the Fighting Ships

In Roman times ships of war, often known as galleys, had a single square sail, but relied on manpower to get up any significant speed. In sea battles there were two ways of attacking the enemy. One was by means of archers, who were carried on a level above the deck, and the other was to drive a heavy ram, projecting underwater from the prow of the ship, into the enemy vessel, powered by the thrust of oars. Banks of slave

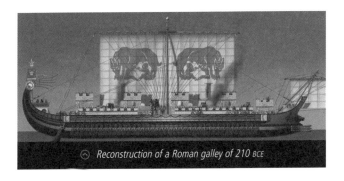

Reconstruction of a Roman galley of 210 BCE

rowers provided the speed required for the attack. The potency of the ship determined its description. Triremes had three banks of oars, quadriremes had four, and quinqueremes five. Each bank of oars required a large number of oarsmen, so that in 315 BCE Roman shipwrights were building galleys with tiers of banks – the very largest of these ships was 400 ft long and 50 ft wide (122 m by 15.2 m) and needed 4,000 oarsmen. As well as the slaves at the oars, the ships carried several hundred soldiers. Just one of these vessels approaching at speed must have been a terrifying sight for the enemy.

As time went on the design of these fighting ships developed until, in the fifteenth century, a type known as a carrack appeared, and the use of oars ceased. A carrack was large, wide and round-hulled, with castles fore and aft; it was also equipped with three or four masts. From the carrack came the galleon, the ship of the English Tudor navy. It was heavily armed, with cannon on gun decks below the main decks firing through hinged ports.

A typical Tudor warship was the *Mary Rose*, mounting 91 guns. Originally built in 1515, she was rebuilt several times, increasing her displacement from 500 to 700 tons. These refits added an extra deck, making her top-heavy and liable to roll steeply in heavy seas. She eventually sank in 1545, off Portsmouth, and was salvaged in 1982, to great archaeological and public interest.

These were heavy fighting ships whose design changed little until the beginning of the nineteenth century. Being reliant solely on sail power they

Henry VIII's purpose-built warship, the Mary Rose. *Ms illustration, c. 1546*

were slow, sluggish to manoeuvre, and forced into fighting in lines astern, using broadsides of fire to attack the enemy.

The last battle between sailing ships was the battle of Navarino, in 1827, between allied British, French and Russian fleets and the Ottoman Turks.

The First Steamers

The development of the steam engine radically changed travelling by sea. It was, as is often the case, the kind of quantum leap that is made by the simultaneous discovery of two things: the technique of boring a large cannon smooth and straight; and the use of that same technique to bore out efficient cylinders for steam engines. Thus the motive power and the armament were set fair for a radical change in the war at sea.

At the beginning of the nineteenth century the British fleet was the finest in the world, and the Lords of the Admiralty wanted it to remain as it was by stating that 'Their lordships feel it is their bounden duty to discourage to the utmost of their ability the employment of steam vessels, as they consider that the introduction of steam is calculated to strike a fatal blow at the naval supremacy of the Empire.'

But the writing was on the wall, as the French and the Americans were already building steam-powered ships, propelled by paddles. In America Robert Fulton built the first of these, equipped with thirty guns and capable of a speed of six knots. However, John Ericsson, a Swedish engineer, invented both a screw propeller that worked and an engine to drive it. He sold the designs to the US Navy and, in 1842, the USS *Princeton* became the world's first screw-propelled steam warship.

In 1859 the French launched the *Gloire*, the first seagoing armoured warship, built of wood with a covering of iron plate. The following year the British responded with the iron-hulled HMS *Warrior*. In the next decade HMS *Devastation* was built with turrets and no sails. The 'ironclads' had arrived, many of them still combining steam with sail power – in the late 1860s USS *Wampanoag*, a steam and sail cruiser, was capable of 18 knots.

These steam ships were powered by engines similar to those used in railway locomotives, until 2 August 1894, when the British-built *Turbinia* was launched. It was equipped with a radical new type of power plant, a steam turbine developed by

Charles Parsons, and reached a speed of 34.5 knots, faster than any of the destroyers of the time. No longer were warships lumbering brutes that needed to catch the wind.

The steam turbine-powered Turbinia *in 1897*

A turbine is a development of the windmill method of producing power. In a turbine a series of vanes on a common shaft are set inside a tube. Hot gases are drawn through the tube opening and expelled, having had their heat energy absorbed in rotating the shaft that is connected to a form of propulsion – wheels in the case of a train, and a screw in a ship.

In a steam turbine the steam produced by a conventional boiler is sucked through the tube and expelled at the rear end. In an aircraft the expelling gases propel the plane along.

The Atlantic Crossing

The Vikings were the first Europeans we know of to make the 1,800-mile journey from Iceland across the Atlantic Ocean to North America, where they built settlements. Led by Eric the Red, they arrived on the north-eastern coast of North America in around 986 CE, but only stayed for a short time. They also established a settlement in Greenland that appears to have lasted until the late 1400s.

> The Vikings crossed the ocean to North America in *knorrs*, with room for around thirty-five passengers and cargo. They were long and narrow, with one mast and a large, square sail. As we have seen, they were also shallow in depth so that the Vikings could take them up river, travelling far inland.

The next serious exploration of the Atlantic route started on 3 August 1492 when Christopher Columbus – an Italian by birth, who was financed by King Ferdinand and Queen Isabella of Spain to find a sea route to the East – left Palos de la Frontera in Spain with three ships; one carrack, *Santa Maria* (nicknamed *Gallega*, the Galician), and two smaller caravels, *Pinta* (the Painted) and *Santa Clara* (nicknamed *Niña*). Columbus first sailed to the Canary Islands, off the north-west coast of Africa, where he restocked with provisions and made repairs. Then, on 6 September, he commenced what turned out to be a five-week voyage across the ocean, first sighting land on 12 October 1492. He had reached the Bahamas.

> The voyage of Columbus was not to discover America but to seek a direct route to the Far East by travelling west. The nearest he got to the Americas we know, was by landing, on his last voyage, on the coast of South America.

Columbus opened the floodgates, and Europe's maritime powers now competed with each other to lay claim to the Americas. Conquest, trade and immigration opened up the Atlantic sea routes.

The Pilgrim Fathers made their way to North America to escape religious persecution in England, with a hundred and two men, women and children travelling in the cramped quarters of the 113 ft (34.4 m)-long *Mayflower*, a small merchant ship.

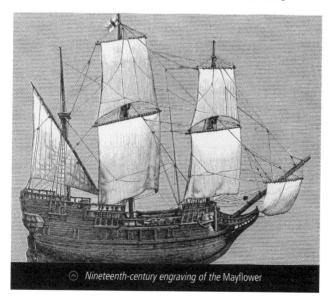

Nineteenth-century engraving of the Mayflower

The party departed from Southampton in September 1620. Initially, the trip went smoothly, but under way they were met with strong winds and storms. One passenger, John Howland, was washed overboard in a storm but caught a rope and was rescued. One crew member and one passenger died before they reached land. A child was born at sea, and named Oceanus. Land was sighted in November, after over two months at sea. They had reached Cape Cod, Massachusetts.

The slave trade

Before long, more and more Atlantic crossings were made, most notoriously to service the burgeoning slave trade. The first report of the use of African slaves in the New World was in 1502; from 1640 to 1680 there was a great traffic of African slaves to the British Caribbean to work in the sugar plantations.

The route taken by the slave ships was known as the Middle Passage. It was so named because the route taken by the ships was triangular. The three legs of the voyages were: England to West Africa to collect the slaves; West Africa to North America to deliver the slaves, and collect the product of the plantations; and North America back to England to deliver these goods.

The death rate was high, although the mortality rate decreased over the history of the slave trade, primarily because the journey took less and less time. In the eighteenth century many slave voyages took at least two and a half months. In the nineteenth century two months appears to have been the maximum length of the voyage, and many voyages were far shorter. However, the number of African deaths directly attributable to the Middle Passage voyage is estimated at up to two million; and evidence suggests that the total number of African deaths en route, directly attributable to the institution of slavery, from 1500 to 1900, was up to four million people.

Cutting the journey time

Once the steam engine became a practical proposition, steam-powered ships were able to cut the time taken to reach the New World from the Old. The first ship of this type to cross the Atlantic was the *Sirius*, in 1838.

The table on the following page shows the increasing speed with which the Old and the New Worlds were brought closer together.

The 700 ton, 320 hp Sirius combined steam paddles and sails

Date	Ship	Notes	Time	Av. speed
1838	*Sirius*	Paddle wheels, rigged for sail, 700 tons, 600 hp	18 days 16 hrs	8.03 knots
1840	*Britannia*	Paddle wheels, rigged for sail, 740 hp, 1,154 tons. The first scheduled transatlantic service	12 days 10 hrs	10.00 knots
1843	*Great Western*	Paddle wheels, rigged for sail. 2,300 tons, 750 hp. Designed by Isambard Kingdom Brunel	12 days 18 hrs	10.03 knots
1863	*Scotia*	Paddle wheels, rigged for sail, 4,000 tons	8 days 3 hrs	14.46 knots
1876	*Britannic*	Steamer, rigged for sail	7 days 13 hrs	15.43 knots
1889	*City of Paris*	Steamer, 28,000 hp, 17,270 tons	5 days 19 hrs	20.01 knots
1900	*Deutschland*	Steamer, 16,502 tons	5 days 15 hrs	22.42 knots
1909	*Mauritania*	Steamer, 90,000 hp, 31,938 tons	4 days 10 hrs	26.06 knots
1936	*Queen Mary*	Steamer, 200,000 hp, 81,961 tons	4 days 0 hrs	30.14 knots
1952	*United States*	Steamer, 248,000 hp, 47,300 tons	3 days 10 hrs	34.51 knots

By the 1950s the age of elegant liners transporting passengers across the Atlantic was over. The jet age had arrived; the liners became cruise ships, and have recently been replaced by floating apartment blocks.

The record for the fastest passenger ship crossing of the Atlantic was held by the liner *United States* (see table above), but in 1986 Sir Richard Branson's powerboat beat that by reaching Bishop's Rock, off the south-west coast of England, 3 days 8 hours 31 minutes after leaving New York.

Rowing across

Over the years various means have been used to cross the Atlantic. The first attempt to row across was in 1898, when two Norwegian fishermen, George Harbo and Frank Samuelson, decided to row from New York to the Isles of Scilly in an eighteen-foot whaling boat. Although they capsized twice on the way, they eventually arrived at the Isles of Scilly fifty-five days later. This record for a west to east crossing of the Atlantic still stands.

The first men in the twentieth century to row the Atlantic were two British paratroopers, John Ridgway and Sir Chay Blyth. They set sail from Cape Cod in June 1966 aboard a 20-foot (6.1 m) dory named *English Rose III*. They arrived ninety-one days later on the Aran Isles, off the coast of Ireland.

The first single-handed crossing was made by Briton Tom McClean, who left Newfoundland in May 1969 in a 20 ft (6.7 m) open rowing boat. Seventy-one days later he landed at Blacksod Bay, in Ireland. In 1982 he again rowed across the Atlantic, this time in a boat measuring just 9 ft 9 ins (3 m) long. It was a world record for crossing the Atlantic in the smallest vessel. In all, he made four single-handed trips, the next in a boat just 7 ft long (2 m). Finally, in 1987, he rowed from Newfoundland to Ireland in just fifty-five days, creating a new record.

A number of other attempts have been were made in more recent years and, at present, the absolute rowing record for an Atlantic crossing is currently held by a French team of eleven which, in 1992, rowed from the Canary Islands to Martinique, in the West Indies, in just thirty-five days.

The Australia Route

Popular histories sometimes state that Australia was discovered by Captain James Cook, of the Royal Navy, in 1770. In fact, the earliest visits to the continent were made in the seventeenth century by Dutch explorers, with the first rough map drawn by Joan Belaeu in 1659.

> Regular sailings between England and Australia began with the First Fleet – the transportation of convicts to the penal settlement in New South Wales. It sailed in May 1787 and consisted of eleven vessels carrying about 1,400 people and stores. The journey took 252 days and travelled for more than 15,000 miles without losing a ship. The route, following the trade winds, began with an Atlantic voyage to Rio de Janeiro. It then proceeded to Australia via the Cape of Good Hope, South Africa, and reached Botany Bay after sailing around Tasmania, finally disembarking in January 1788.

After gold was discovered in 1851, migrants from many countries made the arduous sea voyage to the new British colony. By now the journey was not made solely under sail, for the steam engine had arrived as auxiliary power to the sails.

The route they took at first was from England, down the east Atlantic Ocean to the Equator, crossing at about the position of the Saint Peter and Paul Rocks, around 20° West. A good sailing time for the 3,275 miles (5,271 km) to this point would have been around twenty-one days; however, an unlucky ship could spend an additional three weeks crossing the doldrums – this is a region of calm winds, slightly north of the equator and between the two belts of trade winds, which meet there and neutralize each other, the trade winds being north-easterly winds north of the equator, and south-easterly south of the equator. After this the route then ran south, through the western South Atlantic, following the natural circulation of winds and currents, passing close to Trinidad, then curving south-east past Tristan da Cunha. The route crossed the Greenwich meridian at about 40° South, into the 'Roaring Forties', after about 6,500 miles had been sailed from Plymouth. A good time for the entire run would have been about sixty-five days.

The Suez Canal, in Egypt, opened in 1869 and made a shorter journey possible but, as the ships of the time used sails as well as steam power, they preferred to harness the winds of the 'Roaring Forties'. Only in the early twentieth century was auxiliary sail power no longer needed, and Suez became the standard route from Europe to Australia.

Round the World Sailing Races

The first person to attempt a high-speed sea journey round the world was Francis Chichester, who set himself the goal of beating a 'fast' clipper-ship passage of one hundred days to Sydney. He set off from England in 1966, and completed the run to Sydney in 107 days. He returned via Cape Horn in 119 days. Then Robin Knox-Johnston sailed single-handedly, non-stop, around the world, leaving on 14 June 1968 and returning on 22 April 1969.

In March 2005, Bruno Peyron and crew on the catamaran *Orange II* set a world record for a circumnavigation of 50 days, 16 hrs, 20 mins and 4 secs, and in 2005 Briton Ellen MacArthur set another world record for a single-handed non-stop circumnavigation in the trimaran *B&Q/Castorama*. Her time of 71 days, 14 hrs, 18 mins and 33 secs is the fastest ever circumnavigation of the world by a single-hander. This record still leaves MacArthur as the fastest female singlehanded circumnavigator, but in 2008 Francis Joyon beat the single-handed record in the trimaran *IDEC* with a time of 57 days, 13 hrs, 34 mins and 6 secs.

The great sailing races

The sea provides many ways of creating records. They can be events against the clock, long distance events, or sprints over a fixed distance. For speed sailing over a 500-metre course the current World Sailing Speed Record Council records are:

Class	Speed	Winner	Date
C Class (women)	17.38 knots	Jean Daddo (Australia)	1993
A Class (women)	27.70 knots	Alison Shreeve (Australia)	2005
10m2 (women)	41.25 knots	Karin Jaggi (Switzerland)	2005
Kite Board (women)	42.35 knots	Sjouke Bradenkamp (South Africa)	2007
D Class	46.88 knots	Alain Thebault (France)	2008
A class (men)	43.55 knots	Russell Long (USA)	1992
B Class	44.65 knots	Simon McKeon (Australia)	1993
C Class (men)	46.52 knots	Simon McKeon (Australia)	1993
10m2 (men)	49.09 knots	Antoine Albeau (France)	2008
Kite Board	50.57 knots (outright record)	Alexandre Caizergues (France)	2008

High Speed on the Water – Power Boats

Once powerful racing engines became available, speed enthusiasts were fitting them into boats driven by a propeller and racing them. The following table shows how speeds rapidly increased:

Boat	Driver	Location	Speed	Year
Hydrodome IV	Casey Baldwin	Bras d'Or Lake, Canada	70.86 mph	1919
Miss America VII	Gar Wood	Miami Beach, USA	93.12 mph	1929
Bluebird K4	Malcolm Campbell	Coniston Water, UK	141.74 mph	1939
Slo-Mo-Shun IV	Stanley Sayres	Lake Washington, USA	160.32 mph	1950
Miss Thriftaway	Bill Muncey	Lake Washington, USA	192.00 mph	1960
Miss Budweiser	Dave Villwock	Thermalito Afterbay, USA	220.49 mph	2004

But powering a boat by a conventional engine and propeller was not good enough for the speed seekers. Strapping a jet engine to the back of his boat, Donald Campbell – son of Sir Malcolm Campbell, driver of the famous *Bluebird* cars and boat – took up the challenge to take both the water speed and the land speed records. In doing so he broke the record a number of times, only to lose his life on Coniston Water, in the

⊙ *Model of the* Spirit of Australia, *which set the record in in 1978*

Lake District, in a further attempt in January 1967, after he had reached a speed of 315 mph.

Boat	Driver	Location	Speed	Year
Bluebird K7	Donald Campbell	Ullswater, UK	202.32 mph	1955
Bluebird K7	Donald Campbell	Lake Mead, USA	216.20 mph	1955
Bluebird K7	Donald Campbell	Coniston Water, UK	225.63 mph	1956
Bluebird K7	Donald Campbell	Coniston Water, UK	239.07 mph	1957
Bluebird K7	Donald Campbell	Coniston Water, UK	248.62 mph	1958
Bluebird K7	Donald Campbell	Coniston Water, UK	260.35 mph	1959
Bluebird K7	Donald Campbell	Lake Dumbleyung, Aust.	276.33 mph	1964
Hustler	Lee A Taylor	Lake Guntersville, USA	285.21 mph	1967
Spirit of Australia	Ken Warby	Bowering Dam, Aust.	288.60 mph	1977
Spirit of Australia	Ken Warby	Bowering Dam, Aust.	317.60 mph	1978

Ground Effect Vehicles

Machines that make use of what is known as 'ground effect' rely on being supported on a cushion of air between itself and either water or land. In 1959 the first crossing of the Channel was made by the surface-skimming hovercraft designed by Christopher Cockerell.

Christopher Cockerell used simple experiments, involving a vacuum cleaner motor and two cylindrical cans, to create his unique peripheral jet system, the key to his hovercraft invention, which was patented as the 'hovercraft principle'. He proved the workable principle of a vehicle suspended on a cushion of air blown out under pressure, making the vehicle easily mobile over most surfaces. The supporting air cushion would enable it to operate over soft mud, water, marshes and swamps, as well as on firm ground.

There had been a number of earlier experiments with this type of air cushion vehicle dating back as far as 1915, when an officer in the Austro-Hungarian navy built and tested a fast torpedo boat, where warm air was blown under the hull to create a cushion. Further projects of ground effect systems were carried out in Finland, Russia and America, but Cockerell was the first to design and build a commercially viable method of transport using this concept.

A regular Hoverspeed service ran from Dover to Calais from 1968 until 2000, closing due to increased competition from ferries and the Channel Tunnel. High-speed crossings are now made using catamarans and hydrofoils (boats with wing-like foils mounted on struts below the hull).

A different approach to using ground effect was developed, in great secrecy, in the 1960s by Soviet engineer Rostislav Alexeev, under the general name of *ekranoplan*. An *ekranoplan* has the appearance of an aircraft with short stubby wings and a tail, but has a battery of engines on the body at the front. These craft are very fast over water or land, skimming only a few feet above the surface. One, known as the 'Caspian Sea Monster', and powered by eight engines, was 328 feet (100 m) long, weighed 540 tons (549 tonnes), and could travel at 250 mph (402 kmph).

A fundamental problem with *ekranoplans* was that, while they were very good at travelling in straight lines over clear stretches of land or water, when changing direction they could only do so by taking gently-banking, wide-radius turns.

> Chapter 3
Speed Through the Air: the Early Days

Man's Desire to Fly

Early Man spent much of his time searching for food, and must have looked up with wonder at the flying creatures above him. Animals on the ground he could cope with – after a quick, co-ordinated chase and the possible use of a club or a convenient rock. Birds were beyond his reach. If only he could emulate their mobility and speed, he would have a way of stalking his prey from above, with the added element of surprise.

Our fascination with flight also seems to have deep psychological and religious resonances. In many cultures the heavens are the realm of gods and spirits who are free from earthly suffering and want. By harnessing the winds men sought to share and experience something of this freedom, both as undiluted pleasure and then to more practical effect.

The earliest primitive flying machines were kites, first used in China over three thousand years ago. These were bamboo frames covered in silk or paper – commodities that were well known in China at that time – to catch the wind, .

The first man-carrying aircraft took the form of a kite whose construction was based on a primitive aerofoil. The Chinese used kites for military purposes.

History records that they were of considerable size and some were powerful enough to carry men up in the air to observe enemy movements. Others were used to scatter propaganda leaflets over hostile forces.

> An aerofoil is the shape of a wing section that, when moved through air, creates low pressure on its underside and high pressure on its topside, thus creating 'lift'.
>
> In the National Aeronautics and Space Museum in Washington, D.C., there is a plaque that states, 'the earliest aircraft are the kites and missiles of China'.

While the Chinese were already using kites on a regular basis two thousand years ago, in Europe the technology was not widely known or understood. The only experiments were carried out by intrepid men, often priests, who tried to fly in emulation of the fabled Icarus, by covering themselves with feathers and jumping off the tops of church towers. The results were inevitable, usually ending in the church's graveyard. Given the lack of knowledge of aerodynamics, all these attempts were doomed to failure. Firstly, any lift that might have been created by the 'wings' was unable to support the weight of the 'pilot'. Secondly, these pioneers failed to understand the need for a tail to aid stability. In fact, it took aviators until the late nineteenth and early twentieth centuries to appreciate the need for a rear stabiliser in their flying machines.

The first serious attempts to come to grips with the technology of flight were made by Leonardo da Vinci (1452–1519). Not only was Leonardo a brilliant artist, but he was also a serious inventor and designer of many things, including diving suits, parachutes and flying machines. One of his designs was for what is known as an *ornithopter*, a machine that imitates the flight of a bird. A man was to stand in a wooden boat-shaped hull and manually operate a pair of flapping wings. His surviving plans show that he had appreciated some of the basic requirements of flight, but no attempt appears to have been made to put the design into practice. These and other proposals for flight were doomed from the start, as the technology of the time dictated that the constructions had to be largely of wood and the power provided by a man. It was almost six hundred years before a practical man-powered plane flew across the English Channel.

Balloons

The eighteenth century saw the beginnings of more practical forms of flight – specifically, the use of balloons filled with hot air or hydrogen. Such devices were dangerous, as the gondola slung beneath the balloon had to contain the fire that provided the hot air, and hence the lift, thus the first beings to ascend were not human, but a cock, a duck and a sheep. A few months after this first test a man was carried aloft in the balloon.

> In 1783 the first men ascended into the skies in a balloon designed by the Montgolfier brothers in Paris. They had realised that smoke from a chimney floated upwards and if this could be trapped in a balloon it could provide enough lift to raise a man in a gondola beneath the balloon.

The first crossing, by air, of the English Channel was made in January 1785 by the Frenchman Pierre Blanchard, together with Dr John Jeffries, an

The Montgolfier balloon taking off from the Bois de Boulogne in 1783

American serving in the British Army. They travelled from Dover to a forest near Calais in a time of two and a half hours. This groundbreaking journey was not without its problems as the gallant crew had to jettison everything, including their clothes, in order to prevent ditching in the sea. It was not a particularly auspicious opening to the age of air travel, but it was a notable first, since not only was this the first passenger-carrying aerial crossing of the Channel, a significant distance of twenty-two miles, but Blanchard also carried the first packet of letters by air.

The first attempts to control the flight of these lighter-than-air machines involved the use of paddles as a means of propulsion. Many strange and wonderful attempts at steering were proposed. An early suggestion was to use flocks of birds, driven like horses, attached to an elongated balloon. Various forms of paddle-driven balloons were also tried and, eventually, simple man-operated propellers were used, but an efficient means of propulsion took some time to appear.

In the same century, serious attempts were made to build heavier-than-air flying machines, and a number of successful man-carrying gliders were built. Despite the success of these, the race was on to provide suitable forms of motive power to make flying machines a practical proposition. Lightweight steam engines were tried, and even steam turbines, but it was the building of lightweight internal combustion engines that would provide the power to propel heavier-than-air machines.

Dirigibles and Airships

As balloons were at the mercy of the wind there was a need to devise a way of overcoming the problem. In 1852 Henri Giffard built a balloon whose gas bag was filled with coal gas. Beneath it was hung a platform with a simple steam engine driving a propeller. This was the first *dirigible*.

An engine that was light enough and powerful enough to propel a dirigible was not perfected until 1883, when a dirigible powered by a $1\frac{1}{3}$ hp motor flew in Paris.

Once it became possible to steer a dirigible to go, more or less, where its pilot wished then technology developed apace and rigid-framed *airships* were built. The size of airships rose dramatically and one of the largest of the time, the German L-59, built in 1917, was 225 m (740 ft) long and contained 90,000 cubic metres (2,400,000 cubic feet) of hydrogen gas.

This gas was stored in balloons inside the ship's rigid aluminium frame. Hydrogen gas had, by this time, replaced hot air for keeping the machine in the air, but it was still potentially dangerous as it was highly inflammable. This proved to be a serious problem until the non-flammable gas, helium, could be produced in sufficient quantities to replace hydrogen. In the 1930s the production of helium was controlled by the United States, which banned its sale to Nazi Germany. This, as will be seen later, was to have dramatic consequences.

> By 1916, during the First World War, rigid-framed airships called *Zeppelin*, after their designer Count Ferdinand von Zeppelin, had become a potent weapon of war. German bombing raids across the North Sea, using *Zeppelin*, on London and the English heartland, became commonplace.
>
>
>
> ⊙ *Zeppelin III in flight*

After the First World War, the British airship industry was able to take advantage, as part of the reparations process imposed by the Treaty of Versailles, of technology developed by Germany. The first result of the impetus this gave to the industry was the building of the airships R33 and R34. The second of these made the first trans-Atlantic flight, from England to the USA, in 1919. The outward and return flights covered 7,000 miles (11,200 km) with a total flying time of 204 hours, at an average speed of 34 mph (55 kmph).

> The R100 and R101 were designed as long-range, passenger-carrying airships. R100 was designed by Barnes Wallis, inventor of the dam-busting bouncing bomb in the Second World War, and was privately built. It carried one hundred passengers and ten tons of mail. In 1930 R100 made the first aerial voyage from England to Canada in fifty-seven hours. The R101 was then designed and built under government control, and carried fifty passengers.

Airships were becoming a matter of national pride by the late 1920s and two very significant craft were built in 1929. On 1 October 1930 R101 took off from Cardington, Bedfordshire, where it was built, for Egypt and then India. On board, among a number of government officials, were the Secretary of State for Air and the Director of Civil Aviation. At two o'clock the next morning R101 was flying low over Beauvais in France when it hit the ground and burst into flames. All but four of those on board were killed. One additional casualty was the British airship industry. No more airships were designed or built for fifty years, by which time the highly inflammable hydrogen was no longer in use and much more sophisticated methods of construction were employed. The R100 was sold for scrap for £450 in 1931.

Airship design and construction continued in the USA through the 1930s, and significant advances were made in Germany where supremacy in the air had political, as well as technological, significance. Two great airships emerged during that time. The first was *Graf Zeppelin*, built in 1931 and put into service carrying mail across the South Atlantic. This service alternated with seaplanes catapulted from liners, week by week, continuing until the beginning of the Second World War when the *Graf Zeppelin* and other airships were broken up.

The other notable airship was *Hindenburg*, twice as large as *Graf Zeppelin*. It cruised at 83 mph (134 kmph) against *Graf*'s 75 mph (121 kmph) and, within the hull, it provided luxurious space for fifty passengers, each accommodated in double cabins with facilities found on ocean liners. There was even a lounge equipped with a grand piano. After a few proving flights to South America regular trips across the North Atlantic were made in 1936.

The *Hindenburg* disaster closed the book on airships for long-range passenger carrying. Today, even with the latest technology, it would be of little use attempting to offer trans-Atlantic 'cruises' by airship. They were never intended for speed and they could not fly above the weather. In any case, people needing to travel long distances as quickly as possible soon gave up sea travel once the jet airliner appeared. Modern airships are still in use all over the world, but they are small and largely used for publicity and geological surveys, or by navies for anti-submarine patrols. Once again, a good idea has been bypassed by accelerating technology in other areas.

The Hindenburg *bursting into flames at Lakehurst, New Jersey*

> On 6 May 1937, as *Hindenburg* approached its mooring mast at Lakehurst, New Jersey, it exploded. Of the ninety-seven people on board, thirteen passengers and twenty-two crew were killed. Exploding hydrogen – the source of its lift – soon destroyed the giant airship. The cause of the explosion, only recently identified, was highly inflammable coatings on the airship's fabric being ignited by a random spark.

Powered Flight

It was not until the opening years of the twentieth century that aviation success finally came in 1903, when the Wright brothers' *Flyer* took the world into the age of practical flight with a heavier-than-air machine powered by an internal combustion engine. It may not have been anything more than a long hop, but it was a start. The initial launches were made along a simple track, but a weight-powered catapult, to make takeoffs easier, was used for subsequent flights. The success of *Flyer* then encouraged inventors, particularly in France, to build aircraft of similar design.

> Wilbur (1867–1912) and Orville (1871–1948) Wright were bicycle manufacturers who had built and successfully flown gliders at the turn of the twentieth century. They built and flew the first powered aircraft (the *Flyer*) several times at Kitty Hawk, North Carolina, on 17 December 1903. On the third flight it flew 852 ft (260 m) at a height of 10 ft (3 m) in fifty-nine seconds.

Initially taken up by a few enthusiasts, flying began to be seen as a practical method of transport,

The Flyer taking off at Kitty Hawk, North Carolina, 1903

and flying machine displays became popular novelty events. It took time for the establishment to realise air travel was feasible while, over the next few years, speeds and reliability improved. By 1912 a speed of over 62 mph (100 kmph) was achieved, the first seaplane was flown and a height of 3,281 feet (1,000 m) was reached.

Aeroplanes

Even after the early flying machines had demonstrated that flying was a practical proposition, it took time to demonstrate how aeroplanes were going to influence everyone's lives. Similarly, many governments were slow to be persuaded that flying machines could be used in times of war. Many senior military officers were schooled in the era of the cavalry, and were hard to convince of the possibilities of military aeroplanes.

> In Great Britain, the Royal Flying Corps (RFC) was created in 1912. The Royal Aircraft Factory, at Farnborough, provided the majority of the RFC's aircraft. Initially, they were used for reconnaissance purposes and aerial photography, until pilots began arming themselves with pistols and rifles, and dropped grenades.

At the outset of the First World War in 1914, governments were still unsure of the role of the aeroplane in wartime. The British Royal Flying Corps then had sixty-three aircraft, the French mustered a few more, while the Imperial German Army Air Service had two hundred and forty-six aircraft. Across the Atlantic, the American government had created an Aeronautical Division within the Signal Corps of the US Army as early as 1911 and, by 1912, possessed nine aircraft. These and other aircraft were used by the Americans during the Mexican Revolution but they were not a success. However, by the time the USA entered the war in Europe, in 1917, there were around two hundred and fifty aircraft available for training and military use.

In 1914–15 examples of front line aircraft were:

Aircraft	Engine	Span	Top speed	Ceiling	Armament
Rumpler Taube	Mercedes 99 hp	45.83 ft (14 m)	60 mph (97 kmph)	10,000 ft (3,050 m)	None
BE2A	Renault 70 hp	32 ft (10 m)	80 mph (129 kmph)	11,600 ft (3,540 m)	None
Morane-Saulnier Type L	Gnome Rotary 80 hp	36 ft 9 in (11.2 m)	71 mph (114 kmph)	13,100 ft (4,000 m)	1 x Hotchkiss machine gun

The French Morane-Saulnier was the first true fighter aircraft with a forward-firing machine gun. To prevent the bullets destroying the propeller, steel plates were attached to its blades to deflect any bullets that hit it. It was in April 1915 that the first successful air battle was fought and won by a fighter, a Morane-Saulnier Type L.

The tipping point came quickly on the Western Front when the Germans were equipped with a machine that would cause havoc amongst the lightly armed British and French aircraft: the Fokker E.III, a single-seat monoplane with a forward-firing machine gun.

> The Fokker E.III 'Eindecker' was equipped with a machine gun that fired through the propeller arc. It was fitted with an interrupter gear that stopped the gun from firing when a propeller blade was in front of the gun muzzle.

It was the beginning of the 'Fokker Scourge', and the Allies were forced to develop an armed machine that could play the Fokker at its own game. In doing so, what were originally known as 'scouts' eventually developed into potent single-seat fighters. The backbone of the RFC scout force was the S.E.5.a, designed by the Royal Aircraft Factory, and the rotary-engined Sopwith Camel. Equipped with pairs of forward-firing machine guns, these two aircraft were the mounts of RFC 'aces'—fliers such as Major James McCudden (thirty-eight victories), Captain 'Billy' Bishop (twenty-five victories) and Major William Barker (forty-nine victories).

The Imperial German Army Air Service had a similar number of aces. One of them, Theo Ostercamp, flying the very successful Fokker D.VII, recorded thirty-two victories and went on to fly in combat again during the Second World War. The most famous and successful of them all, however, was Manfred Freiherr von Richthofen who, flying the excellent Albatros D.III and Fokker Dr.I triplanes, downed eighty of his opponents before being himself shot down.

By 1918 fighter aircraft had improved in armament, speed and manoeuvrability, and the best at that time were:

Aircraft	Engine	Span	Top speed	Ceiling	Armament
Fokker Dr.1	Oberursel Rotary 110 hp	23 ft 7 in (7.2 m)	103 mph (166 kmph)	20,000 ft (6,100 m)	2 x Spandau machine guns
Sopwith Camel	100/150 hp Rotary (various manufacturers)	28 ft (2.6 m)	113 mph (182 kmph)	22,000 ft (6,700 m)	2 x Vickers machine guns
Albatross D V	Mercedes 180 or 200 hp	29 ft 8 in (9 m)	116 mph (187 kmph)	20,000 ft (6.100 m)	2 x Spandau machine guns
Fokker D VII	Mercedes 160 hp or BMW 185 hp	29 ft (8.85 m)	117 mph (188 kmph)	22,900 ft (7,000 m)	2 x Spandau machine guns
SE5A	Various 150/200 hp	26 ft 7 in (8.11m)	126 mph (203 kph)	15,000 ft (4,600m)	1 x Lewis machine gun and 1 x Vickers machine gun

In 1914 fighting aircraft carried light bombs and even dropped hand grenades but, by 1918, large bombers were being used by both sides:

Aircraft	Engines	Span	Top speed	Ceiling	Bomb load
Gotha G V	2 x 260 hp Mercedes	77 ft 9 in (23.7 m)	87.5 mph (140.8 kmph)	21,320 ft (6,457 m)	1,100 lb (500 kg)
Handley Page 0/400	2 x Rolls-Royce or Sunbeams 300+ hp	100 ft (30.5 m)	97.5 mph (157 kmph)	8,500 ft (2,000 m)	2,000 lb (910 kg)
Vickers Vimy	2 x Sunbeam or Rolls-Royce 360 hp (max)	68 ft (21 m)	103 mph (166 kmph)	8,000 ft (2,440 m)	2,000 lb (910 kg)
Friedrichshafen G.III	2 x Mercedes 260 hp	77 ft 9 in (23.7 m)	84.5 mph (136 kmph)	15,000 ft (4,600 m)	3,300 lb (1,500 kg)

After the war many of the heavy bombers were converted to carry passengers.

The Schneider Trophy races

As a test of aircraft speeds the Schneider Trophy competition, named after a prize cup donated by Jacques Schneider, the son of a wealthy French armaments manufacturer, was first contested in 1913. Although the event was called a race, it was in fact more of a time trial or speed contest, similar to a hill climb for racing and sports cars, where the competitors attempted the course one after the other, each attempting to complete it in the shortest time.

> The Schneider Trophy was originally intended as an annual competition, but it eventually became a bi-annual event. The nationality of the winning aircraft determined the venue for the next contest. A winner of three successive competitions was allowed to take permanent possession of the ornate trophy.

The competing aircraft had to be capable of landing on and taking off from water, and the test consisted of flying a specified number of laps over a triangular course, the distance covered to be approximately one hundred and fifty sea miles. The winner was the aircraft that covered the course at the highest average speed. Entry was restricted to seaplanes in order to get around the need for the very long runways – which were not available on the grass airfields of the day – that wheeled high-speed aircraft, with high wing loading, would need in order to take off. (Seaplanes, or floatplanes, have their wheels replaced by a pair of floats.)

The choice of venue offered an additional advantage to the organisers since, by holding the event close to a coastline, large crowds could see and hear what was going on.

> Wing loading is the indicator of how 'heavy' an aircraft is. It is not the actual weight, but the loaded weight of the aircraft divided by the area of the wing.
>
> Wings generate lift owing to the motion of air over the wing surface. Larger wings move more air, so an aircraft with a large wing area relative to its mass (that is, with a low wing loading) will have more lift at any given speed.
>
> An aircraft with lower wing loading can take-off and land at a lower speed (or be able to take off with a greater load). It can also turn faster. With a high wing loading, landing and take-off speeds are higher. High wing loading also decreases maneuverability.

It is not surprising that the first Schneider Trophy race was held in Monaco, and won by a French pilot flying a French aircraft – a Deperdussin float plane powered by a 160 hp Gnome rotary engine at a speed of 45.75 mph (74.62 kph). At that time, only ten years after the Wright brothers' first flight, France was the premier country of aviation in the world. Of the four contestants that year, three were French and the fourth hailed from the USA, although flying a French Nieuport. The 1914 race, also held at Monaco, saw the first British entry in the form of a Sopwith Tabloid floatplane that won the Trophy.

Location	Year	Speed mph	Pilot	Country	Aircraft
Monaco	1913	45.75	Maurice Prévost	France	Deperdussin
Monaco	1914	86.78	Howard Pixton	Britain	Sopwith Tabloid
Bournemouth	1919	Race void	Sgt Guido Janello	Italy	Savoia S13
Venice	1920	107.22	Lt Luigi Bologna	Italy	Savoia S12
Venice	1921	117.86	De Briganti	Italy	Macchi M7
Naples	1922	1454.72	Henri Biard	Britain	Supermarine Sea Lion II
Cowes	1923	176.90	Lt Rittenhouse	US	Curtiss CR-3
Chesapeake Bay – Baltimore 1925	1925	232.50	Lt J. H. Doolittle	US	Curtiss R3C-2
Hampton Roads – Virginia	1926	246.50	Maj M. de Bernardi	Italy	Macchi M-39
Venice	1927	281.50	Flt Lt S.N. Webster	Britain	Supermarine S-5
Solent	1929	328.00	Flg Off H.R.D. Waghorn	Britain	Supermarine S-6
Solent	1931	340.30	Flt Lt J.N. Boothman	Britain	Supermarine S-6B

The First World War, together with organisational problems, deferred the next staging of the event until 1920, by which time the technology of flight had improved dramatically.

One of the problems that had to be faced by the competing aircraft at each competition resulted from their being powered by ever more powerful engines. At that time variable pitch propellers were unknown, and the effects of torque (or twisting force) caused by huge amounts of power being delivered via a propeller of very coarse pitch, were considerable. Great care had to be taken to prevent an aircraft rotating about the propeller, rather than the other way round. One attempt to combat this problem was made by the Italian firm of Savoia-Marchetti, with their SM 65, also in 1929. The aircraft was powered by a pair of Isotta-Fraschini Asso engines, each of 1,050 hp. One was mounted in front of the intrepid pilot, who occupied a tiny cockpit, and the other behind him. The propellers therefore rotated in opposite directions, the torque of one cancelling out that of the other. Unfortunately the aircraft crashed while on test, killing the pilot, and the design was never used again.

> A variable-pitch propeller has blades that can be rotated around their long axis, in order to change their pitch (the angle at which they cut through the air), thus adapting the propeller to different thrust levels and air speeds.
>
> If the pitch can be set to negative values, the reversible propeller can also create reverse thrust, for braking or going backwards, without the need for changing the direction of shaft revolutions.

In 1927 the British, all-metal, Supermarine S5 won the Trophy, and started the winning streak of a series of exceptional aircraft using Rolls Royce engines.

The Supermarine S6 and S6B that followed the success of the S5 ensured their success by being powered by a new engine known as the Rolls Royce 'R'. This engine, based on the Rolls Royce Buzzard engine of 825 hp, was eventually boosted to give over 2,000 hp in its final racing form. By this time the effect of the enormous torque produced was becoming a serious problem, so the engineers hit on a solution to prevent the aircraft rotating about the propeller, which was still of fixed pitch. As the fuel was stored in the floats, the starboard float was made to contain 400 lb (800 kg) more fuel than its port partner.

These engines were naturally very highly stressed and, in addition, were fuelled by a mixture of ingredients that was just short of being high explosive. The result was that they could be run for no more than five hours between overhauls. In addition, with 1,900 hp available from the engine in the S6, take-off had to be managed with extreme caution, as the 120 mph (193 kmph) take-off speed was perilously close to its stalling speed, and encountering even the smallest of waves at this point could cause serious problems.

For the 1929 competition two S6 and one S5 aircraft were entered by Great Britain. The main opposition came from the Italians, who fielded a pair of the potent Macchi M67s and a Macchi M52R. An S6 won at a speed of 328.629 mph (529 kmph) with the M52R coming second at 284.2 mph (457.37 kmph).

Although the Trophy went permanently into British hands in 1931, when Flt. Lt. J. N. Boothman won at a speed of 380.30 mph (612.03 kmph), the Italians had the last word with their mighty Macchi MC72s. These formidable floatplanes were powered by huge, twenty-four-cylinder, 3,000 hp Fiat AS6 engines. They would have competed in the 1931 event but trouble with the engine precluded this, leaving the field open for the S6B. However, in

1934 one of these machines reached a speed of 440.682 mph (709.808 kmph) over a 3 km course, and established a still-unmatched speed record for floatplanes.

Crossing the Channel by Air

Louis Blériot made the first heavier-than-air flight across the English Channel in a Blériot XI, an aircraft of his own design, powered by a 24 hp engine of doubtful reliability that carried him from France to England in thirty-seven minutes. Blériot left the French coast at Baraques, near Calais, at 4.35 a.m. on 25 July 1909, and at 5.12 a.m. he landed close to Dover Castle, having flown the twenty-four miles at some 300 ft (90 m) above the waves.

Louis Blériot in his monoplane

Crossing the English Channel unaided

Crossing the English Channel without the aid of any mechanical device has become a challenge and a sport. The first swimmer to cover the twenty-two miles from Dover to Calais was Captain Webb, in 1875, in a time of 21 hrs 45 mins. Since then many hundreds of swimmers have followed his example, some actually swimming the distance twice, both out and back, without stopping. The average speed for these attempts is 1 mph (1.6 kmph).

In July 1979 a lightweight aeroplane named *Gossamer Albatross* flew from England to France. It was powered by a propeller driven by pedals operated by the pilot. It had a span of 97.7 ft (30 m) and in flight it weighed 220 lb (100 kg). The flight took 2 hrs 49 mins flying at an average height of 5 ft (1.5 m).

The First Passenger-Carrying Flights: 1918–52

By 1918 aircraft technology had progressed in leaps, from the primitive flying machines that could only just stagger into the air, to large multi-engined bombers that could carry a ton or more of ordnance over long distances. Civilian flying on a commercial level began when a De Havilland light bomber, operated by the AT&T Company, flew the first fare-paying passenger from Hendon in North London to Le Bourget, near Paris. The first regular, daily flight from England to France commenced on 23 August 1918. The cargo on this inaugural flight consisted of Devonshire cream, braces of grouse, newspapers and a passenger in the form of a newspaper reporter for the *London Evening Standard*. The flight took just under three hours.

> In the very early days of civilian flying it was found that ex-Royal Air Force De Havilland light bombers were ideal for carrying mail across England, together with two passengers in a primitive enclosed cabin. In 1919 members of the British government were flown across the English Channel to Paris for the Peace Conference at Versailles.

While this was going on the RAF was also examining the possibilities of flying, in a series of hops, to India. The first flight, via Paris, Rome and Cairo, was made at the end of 1918, and paved the way for exciting new routes to be opened up. However, it was not until 1 May 1919 that post-war civil flying could take place again, and two companies were allowed to operate services between the United Kingdom and the Continent from 15 June that year.

In 1920 the London Terminal Aerodrome was opened at Croydon, south of London, with three British companies and several French airlines beginning cross-Channel services. Lack of support from the government caused the British companies to pull out in 1921, leaving the new field wide open to European carriers backed by governments eager to open up routes between the UK and the Continent.

By 1920 a London–Paris service was up and running, forming the first leg of regular flights to the Middle East and the Far East. Part of the journey consisted of a train journey from Basle, in Switzerland, to Genoa, in Italy, before picking up a flight to Cairo. This overland section was caused by Italian government hostility to foreigners flying in Italian air space.

In 1931 Imperial Airways introduced what many people believe to be the most elegant aircraft of the time. This was the four-engined biplane Handley Page HP 42 series. These magnificent and sedate biplanes cruised at 95–105 mph (150–170 kmph), had a range of 500 miles (805 km), and were fitted out inside in a style reminiscent of a Pullman car. The seats were fitted with pneumatic cushions and the walls and doors were panelled in mahogany. There were silk curtains at the windows, shaded reading lamps and carpeted floors. The aircraft were used on London–Paris flights, and also on long distance routes to the Middle East.

Imperial Airways aircraft Hanno, *awaiting departure in Palestine, 1935*

Opening up the World – the Birth of Long-Haul Flights: 1918–49

Once the First World War was over the desire to open air routes set aviators flying in all directions across the globe. Hopping across Europe was relatively easy, since the navigational skills required were little more than those required by someone driving off on holiday—if you were lost you could usually land and ask the way. But navigating across large tracts of ocean was another matter entirely. Anyone who dared to attempt a crossing of the two great oceans of the world, the Atlantic and the Pacific, needed at first to be able to steer a course by the stars, provided the visibility was good enough. In addition, they would need an aircraft that had sufficient fuel, and could fly reliably for many hours without breaking down.

The weather over the north Atlantic has always been treacherous but, nevertheless, many attempts were made once the fighting in Europe was over. Aircraft were plentiful (although not necessarily suitable for such a hazardous journey), as were the pilots willing to launch themselves into the unknown. In 1919, Alcock and Brown made the first successful trans-Atlantic flight, from Newfoundland to an Irish bog, in a converted Vimy bomber, a popular mount for these early long distance flights. The flight took sixteen uncomfortable hours to cover the 1,900 miles of the journey. For their achievement both aviators were knighted and presented with a £10,000 prize.

Two weeks before Alcock and Brown's famous flight, the first trans-Atlantic flight had been made by a US Navy flying boat, commanded by Lt. Commander Albert Cushing Read, who flew from Naval Air Station Rockaway, New York, to Plymouth, England, with a crew of five. The journey took twenty-three days, with six stops along the way. The flight won no prizes, however, as it was not non-stop.

Many more flights were attempted, but the failure rate was high, with crews and their aircraft being lost without trace. Among the successful flights was that of a British airship, the R34, which, shortly after Alcock and Brown's flight, flew from Great Britain to the USA and back in 204 hrs.

> The year 1927 saw the first non-stop solo flight across the Atlantic by Charles Lindbergh, taking thirty-three hours to fly from New York to Paris. The following year German aviators made the first east–west crossing. Aircraft technology was improving and, and before long, several successful crossings were made. These flights that beat the ocean were never financially viable as they were heavily laden with fuel. Carrying fare-paying passengers remained a problem that was not solved for two decades.

As the transportation of passengers over long distances was not yet a practical proposition, only mail could be carried. This created further problems, as it was only economical to carry mail if it was in quantity. The ratio of fuel to mail was therefore critical. One solution attempted in 1938

was known as the Short-Mayo Composite. It consisted of a small float-equipped mail-plane perched on the back of a Short Empire flying boat. This 'pick-a-back' aircraft, known appropriately as *Mercury*, carried enough fuel for its flight with the mail on board. The lower component, *Maia*, had the task of getting its heavily loaded charge into the air. Once airborne the two aircraft parted company and *Mercury* flew on to its destination. The experiment was successful and the combination carried mail non-stop, both to Alexandria and to the USA. But passengers still had to travel across the Atlantic by liner until early 1939.

Meanwhile, the Germans had been approaching the problem from a different angle. Their method was to catapult a mail-carrying seaplane from a liner 250 miles (400 km) out from New York. Since aircraft travel faster than ships the mail arrived at New York a day before the ship. That was in 1928. The experiment was so successful that it was continued for many years. Then, in 1933, a new approach was made to the problem. This was to use a flying boat and a 'mother ship' stationed halfway along the route. The flying boat landed near to the ship and was hauled on board for refuelling. Then it was transferred to a catapult to continue its journey.

A flying boat differs from a seaplane in having the passenger cabin built on a planing bottom, similar to that of a sailing boat.

> Pan American launched a trans-Atlantic service in 1939 with their new Boeing 314 flying boats. The first flight, from Newfoundland to Southampton, carried nineteen passengers and took just under nineteen hours. Unfortunately the Second World War started and the service was curtailed. However, in 1941, BOAC purchased three of the Boeing flying boats from Pan Am. They flew east–west, making four stops on the way and, on one occasion, flew Winston Churchill from Norfolk, Virginia, to Plymouth. This was the first trans-Atlantic flight by a British Prime Minister. The Boeings remained in service until 1946.

During the war trans-Atlantic flights became commonplace, a feat born of necessity. In the early days of peace, converted bombers were used before specially designed long-haul airliners began to appear. In 1949 the Boeing 377 *Stratocruiser*, developed from the B29 *Superfortress* bomber, came into service. The Atlantic flight was scheduled to be non-stop overnight, but often stopped to refuel at Prestwick, in Scotland, and at Gander, in Newfoundland. The pressurised Stratocruiser carried one hundred passengers at 301 mph (483 kmph) over 4,000 miles (6,760 km).

> In 1987 Richard Branson crossed the Atlantic by hot-air balloon. The *Virgin Atlantic Flyer* was the largest balloon ever made. It had a capacity of 2.3 million cubic feet, and could travel at over 130 mph (209 kph). This feat made history: it was the first time that a balloon had successfully crossed the Atlantic.
>
> In 1991, Sir Richard went even further by crossing the Pacific in a balloon that travelled at speeds in excess of 245 mph (390 kmph), breaking all existing records.

Flying Eastwards: 1918–39

With the ending of the First World War, those countries possessing air power determined to set up commercial flights across the continents. Britain still had an empire, and it was deemed imperative that new and faster ways of communicating with the colonies were found. The two immediate objectives were to open up routes to South Africa and India. The second of these was particularly important as it provided an excellent jumping off ground for later flights to Australia and New Zealand.

On the face of it, a flight to the Cape was straightforward – north to south, down the continent of Africa – and, as early, as the end of 1918, the route had been surveyed by teams from the newly formed Royal Air Force. From Cairo, the route took the aircraft over the Sudan, Kenya, Uganda, Tanganyika, Northern and Southern Rhodesia, and finally South Africa. Over forty supporting landing grounds were established by the end of 1919. Aviation was big news at the time, with intrepid aviators from all over the world constantly attempting to set new records for longer and faster flights, and it was then that the *Daily Mail* newspaper offered a £10,000 prize to the first crew to complete the trip from Cairo to the Cape.

By January 1920 five aircraft and crews were ready to set off from London on the first leg of the journey to South Africa. Four of the aircraft were converted twin-engined bombers, and the fifth was a single-engine De Havilland. Although they were all competing in a race for the *Daily Mail*'s prize, they took off separately, over a period of a month.

Not one of the five original aircraft made it to the Cape, although the crew of a Vickers Vimy bomber was sent a replacement aircraft, enabling them to complete the journey. Unfortunately, since the trip was made in heavily repaired machines, the prize could not be awarded to that entry. Despite the many accidents that befell the aircraft, none of the crews suffered anything more than minor injuries. One might have expected that after so many setbacks and failures those unlucky flyers would disappear into oblivion. But this was not the case; in time, most of them rose to positions of eminence in the aviation world.

Trail Blazers and Record Breakers: 1925–37

Success for Alan Cobham

It was not until 1925 that a further attempt was made by Alan Cobham, one of the original competitors in 1919, to complete the Cairo to Cape flight. This was made under the auspices of the recently formed Imperial Airways, which had learned the lessons of the past and was determined to overcome the obstacles presented by flying over inhospitable terrain. These varied from the steamy swamps of the Sudan, to high altitude airfields, such as Jinja in Uganda, at a height of 4,000 ft (1,220 m). Cobham, a very experienced flyer, was already known as a dedicated publicist of flying. His popular barnstorming air shows brought aviation to people all over Great Britain.

The aircraft to make the flight was carefully chosen and tested by Cobham. The flight was planned meticulously and each of the twenty-five landing sites used on the trip down the continent was accurately surveyed. On board the De Havilland with Cobham were an engineer and a Gaumont–British Picture Corporation cameraman. The cameraman's task was to film the journey with a view to using the record of the flight to attract the public into taking advantage of the service that the airline intended to open.

Having left Croydon on 16 November 1925, Cobham completed the 8,000-mile (12,900 km) journey without serious mishap on 17 February 1926. The return journey to Croydon was completed on 13 March 1926. The trip was completed in 174 flying hours at an average speed of 46 mph (74 kmph). In 1926 Alan Cobham was knighted for his services to aviation.

'Amy, Wonderful Amy'

The 1930s were, in many ways, the golden years of aviation. Flying was adventurous and glamorous, and any trailblazing flight was guaranteed to be instant front page news. But if a flight was made by a woman then the interest and adulation was equivalent to that accorded a modern pop star. Amy Johnson became the darling of the nation with her record-breaking flights, to the extent that a popular songwriting duo of the day immortalised her in song.

Born in Hull in 1903, Amy Johnson took up flying almost by accident following a chance visit to the

London Aeroplane Club in 1928. By the end of 1929 she had not only gained her Private Pilot's Licence (PPL), but also a Ground Engineer's Licence – the first woman to do so. Not content with these achievements, she set her sights on being the first woman to fly the 11,000 miles (17,700 km) from England to Australia. With no finances of her own to purchase an aircraft, she was lucky enough to obtain help from her businessman father and from Lord Wakefield, head of an international oil company. In 1930 she purchased a De Havilland Moth, named *Jason*, and in May that year flew from Croydon to Darwin in nineteen days, at an average speed of 24 mph (37 kmph). Other long-distance flights followed: England–Tokyo in just under nine

⊙ *Amy Johnson in Kalgoorlie, Western Australia*

days in 1931 (28 mph, 45 kmph); England–Cape Town in 4 days 6 hours in 1932 (average speed 59 mph, 94 kmph). Another England–Cape flight in 1936 took 3 days 6 hours, with the return flight taking 4 days 16 hours. This last trip established Amy Johnson as the holder of the out-and-back record time. Her England–Cape record was to stand until 1967, when it was beaten by Sheila Scott.

Amy Johnson's first flight to the Cape had beaten the time taken by Jim Mollison (4 days 17 hours) in the same year by ten hours. The two aviators married and continued record-breaking flights together. The most spectacular of these was an east–west flight across the Atlantic in 1933 in an attempt on a long-distance flight record, the trip taking 39 hours (77 mph, 103 kmph).

When war came Amy Johnson offered her services as a ferry pilot and was lost without trace when the aircraft she was flying crashed in the Thames Estuary in 1941.

Alex Henshaw

Alex Henshaw was one of Britain's great aviators. He began his flying career in 1932 and remained actively involved in aviation until his death in 2007. Having learned to fly he soon took up air racing, eventually winning the King's Cup Air Race in 1938 in his Percival Mew Gull single-seater. It was in this soon-to-be famous aeroplane that he took on the challenge of flying from Gravesend, in England, to Cape Town. The flight involved a number of refuelling stops, the first at Oran (Morocco), then at Gau (Sudan), Libreville (Gabon) and Angola, finally arriving at Cape Town having flown at an average speed of 152 mph (245 kmph). The journey of 6,377 miles (10,361 km) was covered in 39 hrs 28 min, beating the existing record by twenty minutes. After resting in Cape Town for less than a day Henshaw then made an immediate return trip, taking only thirteen minutes longer than the outward flight. That record still stands.

During the war Henshaw distinguished himself as chief pilot at a major aircraft factory that built Supermarine Spitfires. He tested almost 3,000 of the aircraft prior to their being entered into service.

Charles Lindbergh

As we have seen, the great American aviator Charles Lindbergh flew across the Atlantic single-handed, from New York to Paris, in May 1927 – 3,415 miles (5,500 km) in 33 hours at an average speed of 103.5 mph (166.5 kmph). His single-engined aircraft, *The Spirit of St Louis*, had been built specially for the trip by Ryan Airlines of San Diego, California. His grandson Erik Lindbergh repeated this trip seventy-five years later in 2002.

Lindbergh Sr.'s first solo flight across the Atlantic won him the Orteig Prize of US $25,000 and a ticker-tape parade on Fifth Avenue, New York

Charles Lindbergh in the cockpit at Lambert Field, St. Louis

City, on 13 June 1927. He was presented with the Medal of Honor for his historic flight. Lindbergh's public stature, following this flight, was such that he became an important spokesman for aviation. He served on a variety of national and international boards and committees, including the central committee of the National Advisory Committee for Aeronautics in the United States, and served for forty-five years as a technical adviser to Pan American Airways, being instrumental in developing their long-range passenger services.

Amelia Earhart

Amelia Earhart was born in Kansas in 1898, and first flew solo when she was twenty-three years old. It was not, however, until 1928 that she did any serious flying – although not at the controls. This flight was made as a passenger in a trimotor Fokker from New York to London, and gave her the distinction of being the first woman to fly the Atlantic. She spent the next year promoting the cause of women flyers with the backing of the publisher George Putnam, whom she subsequently married. She set speed records and an altitude record for autogyros (see page 141), as well as for a solo flight across the Atlantic to Ireland, the journey taking fifteen hours. After this flight her husband nicknamed her 'Lady Lindy', after record-breaker Charles Lindbergh.

Epic flight followed epic flight. In January 1935 Amelia Earhart became the first person to fly solo across the Pacific Ocean, from Hawaii to California.

In 1936 she flew solo from Newark, New Jersey to Mexico, and also from Los Angeles to Mexico.

She was fêted all over the USA, filling her time, when not flying, with lecture tours and writing. Two volumes of her autobiography were published: *20 Hrs., 40 Min.: Our Flight in the Friendship*, recalling her first trans-Atlantic flight; and *The Fun of It*. By 1937 she came to realise that her flying adventures could not last a lot longer, as there were fewer and fewer long-distance records to break. There was, however, one more record she was determined to attack. This was to be a flight around the world, keeping as close to the Equator as possible. For the attempt she acquired one of the latest products of the Lockheed Company, in the shape of a twin-engined, all-metal airliner. The Electra, named *Lady Lindy*, was specially adapted for the flight, being fitted with special long-range petrol tanks, automatic pilot, and the latest radio equipment.

The aircraft carried a crew of four. There was Amelia herself, a co-pilot and two navigators. Her companions were all highly experienced and the whole journey was meticulously planned. Their first attempt, in March 1937, met with failure. At first all had gone well on the leg from Oakland, California, to Honolulu. But on take-off from Honolulu the Electra groundlooped, was damaged, and was shipped back to California for repair. In May the flight restarted, but in the opposite direction, and with only Fred Noonan, a navigator, accompanying.

Speed Through the Air: the Early Days

Amelia Earhart and the Lockheed L-103E Electra, c. 1937

The flight went as planned down to Brazil, across the south Atlantic to Dakar, across Africa and India, and then to Bangkok. By this time the flight had taken a month, and they were heading for Java and thence Darwin, in Australia. The last section of the journey was in sight and they were aiming to return to the United States on 4 July. From Darwin they flew to New Guinea and, on departing, they headed for Howland Island, some 2,550 miles (4,100 km) away. Units of the US Navy were detached to keep station along the route in order to assist in the aircraft's navigation. But *Lady Lindy* never reached the island. The aircraft and crew were lost in the Pacific, for reasons unknown to this day. Noonan was an experienced navigator, having flown with Pan Am on their proving flights over the Pacific. But this was the longest leg of their flight and the slightest navigational error would have put them miles off course.

Various theories have been put forward over the years as to exactly why and how they disappeared. One of these was that they were deliberately off course on a secret spy flight over Japanese territory in the Caroline Islands. It was thought that in doing so they crashed, were made captive, and possibly executed. The probability, however, is that they ran out of fuel and crashed into the sea in the region of their destination after twenty hours in the air.

The Challenge of the Pacific: 1932–39

By the late nineteen twenties European airlines were setting up passenger and mail air services joining Europe to the Far East, America and Africa. To reach China and Japan, the Americans had a more difficult route to plan as it involved flying across the 7,500 miles (12,000 km) of the Pacific Ocean. Accurate navigation across that ocean posed a serious problem and, additionally, there were no aircraft available with the range necessary to make a non-stop journey. The only route possible was to fly island to island across the Pacific, the first stage being from San Francisco to Hawaii, but even that involved a flight of 2,400 miles (3,700 km) across the ocean.

> For long-distance travel over the ocean, aircraft of the time could not carry sufficient fuel and passengers for the journeys between the islands on the route. Not only that; the islands were rarely large enough to provide runways for land-based aircraft. Large flying boats had to be the answer.

Speed Through the Air: the Early Days

Pan American Airways decided to meet the challenge and use flying boats as they needed no airstrips. What was needed for a regular trans-Pacific service was the establishment of island staging posts with space for workshops, fuel dumps and accommodation for passengers. Survey flights on the route were commissioned to prove that it could be used commercially. Flying boats were ordered and the survey flights began in April 1935, with flights between the staging posts. The final distance of the service from San Francisco to Manila was 8,000 miles (12,900 km) and took 60 hours (5 days); the first passenger service took place in October 1936.

By the end of 1937 a total of over two thousand passengers and a regular 250,000 items of mail had been carried each month over a total of 1,500,000 miles (2,400,000 km). The service operated successfully until that day in 1941 when Pearl Harbor was attacked. It was never to return.

The Sikorsky S-42 flying boat

Scheduled services soon became possible as by 1937 the route extended to the Chinese mainland, taking six days for the flight to Hong Kong from the USA. Macau and Singapore also became destinations for the service. Survey flights established further routes to New Zealand. Using Pearl Harbor as a jumping-off ground, the service to New Zealand was established via Kingman Reef and Pago Pago, and terminated at Auckland. The final proving flight was made by *Pan American Clipper II*, which flew the Hawaii–Kingman Reef route of 1,000 miles (1,610 km) in 7.5 hours (133.3 mph, 215 kmph) and within a day of landing set off on the 1,600-mile (2,600 km) journey to Pago Pago. The final section of the flight that was to take them the 1,850 miles to Auckland, New Zealand, was achieved in 12 hours (152.5 mph, 245 kmph).

West to East by Imperial Airways: 1931–39

In 1938 there were no direct flights, or series of flights, readily available from Great Britain to Australia and New Zealand. The only regular flights to these outposts of the Empire were via America and Pan Am.

A route of sorts joining Europe with the Antipodes had been started early in the 1930s when Imperial Airways began an experimental airmail service that left Croydon bound for Darwin in April 1931. The journey was made in a series of stages, and the mail was initially flown to Karachi, where the cargo was taken over by another aircraft. Unfortunately it crashed in Timor, and the mail was carried the rest of the way to Darwin in an aircraft of Australian National Airlines.

A service by Imperial Airways was opened in December 1935 when a Handley Page HP 42 carried the first regular mail from Croydon to Karachi, where the cargo was offloaded on to two further aircraft en route for Darwin. Two Australian aircraft shared the final leg of the route to Brisbane, arriving two weeks later. When the service became available to passengers as well as mail, the journey from London to Australia took over twelve days and involved thirty-five stops.

Short C-Class flying boats joined the now aging Handley-Pages in 1936. They operated out of the coastal waters off Southampton, and by 1938 twice-weekly services were being flown between England and Australia. The passengers made the journey in the height of comfort, since it was in competition with the sea route, so that it was in effect an aerial cruise. A typical journey from England to Australia would cover 13,000 miles (21,000 km) in nine days by flying boat. On board would be twenty-four passengers, one 'air hostess' and one steward serving the in-flight food and

Qantas Short C-Class Empire flying boat *Coolangata*, c. 1940

drink. Given the flying boat's range of 2,350 miles (3,800 km), and cruising speed of 180 mph (290 kmph), overnight stops were made with the passengers staying in hotels, with time often allowed for sightseeing.

After the war the frills that had made the journey an event were dropped, and the flying boat service was replaced by something rather more practical. In May 1945 an Avro Lancastrian, a converted Lancaster bomber, took off from Hurn carrying nine passengers, heading for Sydney. The service was operated jointly by BOAC (Imperial Airways having become the British Overseas Airways Corporation during the war) and Qantas (Queensland and Northern Territories Aerial Service), and made stopovers at Lydda (today, Israel), Karachi, Ceylon and Learmonth (western Australia). The handover between the airlines took place at Karachi. The nine passengers were accommodated side-by-side, facing outwards, on the starboard side, and there was provision for six sleeping berths. The journey took 66¼ hours, including refuelling stops. The passenger service using Lancastrians lasted until 1947, but continued for the carriage of mail and freight until 1950.

In 1947 Qantas introduced modern aircraft into their fleet in the shape of Lockheed L 749 Constellations. These were one of the new breed of aircraft seating up to a hundred passengers in pressurised comfort. They could fly 5,400 miles (8,700 km) at a cruising speed of 340 mph (547 kmph). These, and the trans-Atlantic Boeing Stratocruisers, were at the edge of technology as far as speed and comfort were concerned. They filled the gap until the first jet liners appeared.

By October 1947 the first Qantas 'Connie' was delivered and in December that year made the first passenger flight, taking ninety-four hours to reach London Heathrow, carrying twenty-nine passengers and having made six stops. The passengers on that first service would have paid the equivalent of nine times the first-class fare on a modern Qantas 747 – but what a difference compared to the noisy, lumbering old Lancastrian, with its passengers seated side-by-side just above the bomb bay! The passengers on that trailblazing flight travelled at 18,000 ft (5,500 m) in pressurised comfort at an incredible, for those days, 300 mph (500 kmph). The actual time spent in the air amounted to fifty-six hours. As well as the passengers and eleven crew, the aircraft carried mail and a ton of food (including five hundred Christmas puddings) for the still-rationed British.

Until 1946 the only way to fly direct from Europe to New Zealand was by flying to the USA and then picking up the Pam Am flight to Auckland. But in that year regular flights from Sydney to Auckland were linked with BOAC/Qantas flights from London with a service across the Tasman Sea, and the gap was closed.

The first jetliners to fly were the De Havilland Comet (500 mph, 810 kmph) that flew in 1949, followed by the Boeing 707 (622 mph, 1,000 kmph) in 1954, and the Sud Aviation SE-210 Caravelle (575 mph, 925 kmph) in 1955.

Chapter 4
Speed Through the Air: the Age of the Jet

Between the wars the design of military aircraft did not change significantly. In 1939 the RAF was only just beginning to be equipped with fast monoplane fighters such as the Hurricane and the Spitfire, and the bomber force was hardly any readier for war. The German *Luftwaffe*, meanwhile, had been developing and gaining combat experience in the Spanish Civil War in the late 1930s. However, the war years of 1939–45 forced the pace of development on both sides. In 1939 typical front-line fighters were:

Aircraft	Engine	Span	Top speed	Ceiling	Armament
Supermarine Spitfire Mk 1	880 hp Rolls-Royce Merlin	36 ft 10 in (11.2m)	362 mph (580 kmph)	34,000 ft (10,400 m)	8 x Browning machine guns
Hawker Hurricane Mk 1	1,030 hp Rolls-Royce Merlin	40 ft (12.2m)	340 mph (545 kmph)	40,000 ft (12,200 m)	8 x Browning machine guns
Messerschmitt Bf 109E	1,550 hp Daimler Benz	32 ft 4 in (9.9m)	350 mph (560 kmph)	35,000 ft (10,700 m)	4 x MG17 machine guns + 1 x 20 mm MG FF

The Spitfire IIA P7350

Speed Through the Air: the Age of the Jet

Front-line bombers of the period were:

Aircraft	Engines	Span	Top speed	Ceiling	Bomb load
Armstrong Whitworth Whitley	2 x 1,145 Rolls-Royce Merlin	84 ft (24.6 m)	230 mph (370 kmph)	26,600 ft (7,900 m)	7,000 lb (3,200 kg)
Vickers Wellington	2 x 1,050 hp Bristol Pegasus	86 ft 2 in (26.3 m)	235 mph (380 kmph)	18,000 ft (5,500 m)	4,500 lb (2,000 kg)
Dornier Do 217	2 x 1,580 hp BMW (typical)	62 ft 4 in (19 m)	320 mph (515 kmph)	24,600 ft (7,500 m)	8,818 lb (4,000 kg)
Heinkel He 111	2 x 1,200 hp Jumo	74 ft 2 in (22.6 m)	258 mph (414 kmph)	25,600 ft (8,000 m)	4,410 lb (2,000 kg)

Aircraft became more efficient in their designated tasks. Thus by 1945 the finest piston-engined and the first jet fighters in service were:

Aircraft	Engine	Span	Top speed	Ceiling	Armament
Supermarine Spitfire Mk XIV	2,005 hp Rolls-Royce Griffon	32 ft 8 in (10.0 m)	448 mph (720 kmph)	44,500 ft (13,600 m)	2 x 20 mm cannon and 2 x 0.5 inch machine guns
Hawker Typhoon	2,1880 hp Napier Sabre	41 ft 7 in (12.7 m)	412 mph (663 kmph)	34,000 ft (10,370 m)	4 x 20 mm cannon plus 250 lb bombs
North American P-51D Mustang	1,590 hp Packard Merlin	37 ft 1 in (11.3 m)	448 mph (720 kmph)	41,900 ft (13,000 m)	6 x ? in machine guns
Focke-Wulf Fw190D	1,776 hp Jumo	33 ft 5 in (10.2 m)	440 mph (708 kmph)	32,800 ft (10,000 m)	1x 30 mm MK108 plus 2 x 20 mm MG 151 and one 1,100 lb bomb
Gloster Meteor	2 x Rolls-Royce Welland turbojets	43 ft 1 in (13.1 m)	410 mph (660 kmph)	34,000 ft (10,370 m)	4 x 20 mm cannon
Messerschmitt Me 262	2 x Jumo turbojets	41 ft (12.5 m)	540 mph (870 kmph)	37,565 ft (11,500 m)	4 x 30 mm MK 108 cannon

Similarly, by 1945 heavy bombers in service were:

Aircraft	Engine	Span	Top speed	Ceiling	Armament
Avro Lancaster Mk X	4 x 1,300 hp Packard Merlin	102 ft (31.1 m)	287 mph (462 kmph)	19,000 ft (5,800 m)	22,000 lb (10,000 kg)
Boeing B-17F	4 x 1,200 hp Wright Cyclone	103 ft 9 in (31.6 m)	295 mph (475 kmph)	36,000 ft (11,000 m)	17,600 lb (8,000 kg)
Heinkel He 177	2 x 2,950 hp Daimler Benz	103 ft (31.4 m)	295 mph (475 kmph)	26,500 ft (8,100 m)	13,200 lb (6,000 kg)

Designing the Jet

In April 1937 a strange new sound was heard to come from the primitive workshops of a company that had been formed twelve months earlier by a junior Royal Air Force officer named Frank Whittle. The company was Power Jets and the sound would eventually be heard all over the world. It came from the very first gas turbine aircraft engine, where burning fuel passes through a turbine, exiting at high pressure and producing thrust. The first jet engine was being tested.

Yet the whole thing was viewed as being of dubious value by the 'experts' in the Air Ministry, who failed to spot a potential war-winner and slowed down any development in the field in 1940 in favour of the production of such 'formidable' (*sic*) aircraft as the Bristol Blenheim and the Armstrong Whitworth Whitley. The first flight by a British jet-propelled aircraft was made by the E.28/39 built by the Gloster Aircraft Company in May 1941. After the historic first flight an onlooker remarked to Whittle, 'It flies!' To which Whittle's reply was 'That's what it was bloody well supposed to do!'

First prototype of the Gloster E.28/39

Frank Whittle (1907–96) had gained a First in Mechanical Engineering at the University of Cambridge, under the auspices of the RAF, which generously allowed him to spend all of six hours a week developing his concept that was to revolutionise air travel in the next decade. This time was given grudgingly by Whittle's superiors who had little faith in a revolutionary device that was the brainchild of a junior officer.

By the autumn of 1941 there were only two Whittle engines in existence and the Ministry, having seen that the device they had sneered at actually worked, made the decision that Whittle's company (Power Jets) should hand over all its expertise to the Rover company, who were excellent at making motor cars, for further development! Very little happened, owing to massive internal squabbles, until the work was finally given to Rolls Royce, who started to work in earnest during 1943.

However, it was not only in England that this line of research was taking place. A German engineer, Hans von Ohain, a research assistant at Göttingen University, had convinced Ernst Heinkel that a propellerless aeroplane was a practical proposition. His first effort, fuelled by hydrogen gas, ran on almost the same date as Whittle's first engine. Strangely enough the two lines of development in England and Germany were proceeding, unknown to each other, in parallel. The only difference was that Whittle's research was allowed to continue, almost under protest, by the government of the day while the Heinkel project had the backing of the RLM (*Reich Lufthart Ministerium*), the German Air Ministry. The German *Luftwaffe* was being expanded at a phenomenal rate in 1936 and Von Ohain had the full cooperation and financial backing of the authorities.

> Hans Joachim Pabst von Ohain (1911–98) was born in Dessau, Germany, and gained a PhD in Physics and Aerodynamics from the University of Göttingen. His engine used a centrifugal compressor and turbine placed very close together, back to back, with the flame cans wrapped around the outside of the assembly. The resulting engine was even larger in diameter than Whittle's, although much shorter along the thrust axis.

The successful running in 1938 of the German jet engine, this time fuelled by petrol, took place only a few months after the Whittle engine, but the first flight of a jet-powered aircraft in Germany was in August 1939. This was the Heinkel He 178, a full two

Messerschmitt Me 262, the world's first jet fighter

years before the Whittle-powered Gloster machine. The Heinkel He 178 and the Gloster/Whittle E28/39 were both designed as research aircraft, and both proved the practicality of jet power. The RLM grasped the concept and gave a go-ahead for further development. As a result the first production jet fighter was the Messerschmitt Me 262, which first flew under full jet power in July 1942. This early German success predates the first flight of Britain's Gloster Meteor, which did not fly until March 1943.

The development of jet propulsion in the USA lagged far behind developments in Europe. In fact, the Americans had never heard of either Whittle or Von Ohain. But one day in 1941, before America entered the war, General 'Hap' Arnold of the US Army Air Corps, who had been sent to England, met Lord Beaverbrook, Churchill's

Minister of Aircraft Production and one of his most intimate advisors. Beaverbrook turned to Arnold and said, 'What would you do if Churchill were hung, with the rest of us in hiding in Scotland or being overrun by the Germans? What would the people in American do? We are against the mightiest army the world has ever seen.' Those present at the meeting agreed that Germany could invade England at any time. The Americans, although not at war, were sympathetic to Britain's situation. Nevertheless they drove a hard bargain.

They were prepared to provide Britan with much needed warplanes, but in return the British would supply the plans for the Whittle turbojet engines. This was agreed to, provided the utmost secrecy was maintained and a strictly limited number of persons would be involved in development of the engine in the USA. Arnold personally inspected the Whittle engine and witnessed the first flight of the Gloster E./28/39. He arranged to have General Electric's Supercharger Division take on the American development of Whittle's prototype and selected Bell Aircraft of Buffalo to work concurrently on an airframe for a jet fighter aircraft.

By September 1941 it was decided that General Electric should build fifteen jet engines based on the Whittle design, and Bell was chosen to build a suitable twin jet engine fighter aircraft. A contract for US $1.6m to build three prototypes was given to Bell for the first aircraft to be made available in May 1942.

Speed Through the Air: the Age of the Jet

The Whittle W2/700 turbojet engine

Gas turbine engines (jets) have been built in various forms. Frank Whittle's first engine, and its derivatives, were of the centrifugal type that cause the flow of air through the compressor to be perpendicular to the axis of rotation.

In an axial compressor, the airflow through the unit travels parallel to the axis of rotation. This gives the engine a slimmer line.

Virtually all jet engines in use today are of the turbofan design, where air first enters through a large fan, powered by the turbine, and thrust is produced by the fan and the compressed burnt gases.

The Jet Fighters

Although the Germans had a two-year lead over the British, the introduction of their jet fighter suffered many delays. In the case of the Messerschmitt Me 262 these were partly due to interference by Hitler, who wanted this formidable machine to be a bomber, whereas the *Luftwaffe* knew that it was just what was needed to combat the daylight raids over Germany by American B-17s and B-24s.

The Messerschmitt jets were eventually introduced operationally in late 1944, but suffered from problems caused by the shortage of materials for the construction of the engines and from a lack of trained pilots. The jet engines proved to be unreliable, due to overheating, and this led to an operational life of just twenty-five hours. When they did go into action they created havoc among the Allied aircraft, scoring a significant number of victories over the bombers and their escorting fighters. However, although their speed was vastly superior to P-51 Mustangs, Hawker Tempests and Typhoons, a surprising number were shot down.

As an example of the panic measures that were introduced in Germany in the last stages of the war, the Heinkel He 162 was designed and built, largely of wood, in a matter of weeks at the end of 1944. Because of the shortage of pilots it was decided, as a last-ditch measure, that members of the Hitler Youth should undergo a crash course learning to fly gliders, and then be put into the cockpit of what turned out to be a very potent fighter. The Heinkel He 162 was a difficult aircraft to fly, even in the hands of an experienced pilot, so how it would have performed if it had entered service flown by a raw teenage recruit doesn't bear thinking about.

Meanwhile the first British jet fighter, the Gloster Meteor, was being developed, but was too late to be used tactically. Only fifty-eight Meteors were in RAF service at the end of the war, whereas the *Luftwaffe* had 1,300 Me 262s, although only a fraction of these were in service.

At the end of the war Whittle's Power Jets Ltd was nationalised and required to hand its designs to Rolls-Royce, and Power Jets Ltd was denied permission to build jet engines. To add insult to injury, the companies that took over Whittle's work then failed to employ the very man who had made it all possible. By these actions, Sir Stafford Cripps, the Minister of Supply at the time, effectively gave the gas turbine technology to other aero engine manufacturers in Britain, as well as to the United States and the Soviet Union, allowing them to develop their own jet-powered aircraft.

Ironically, Whittle had been working on the design of a turbofan engine to power a four-engined bomber capable of bombing Japan. After VJ (Victory over Japan) Day, 15 August 1945, the powers in Whitehall announced that this project was to be halted, as there was no need for such an aircraft. Nobody had the forethought to realise that the transformation of such a design into a long-range transport aircraft would provide exactly what the airlines had been clamouring for for years. This resulted in misguided projects to build huge propeller-powered white elephants, none of which were fit for purpose. And what are many airliners powered by now? Turbofans.

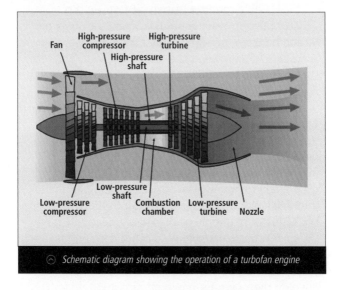

Schematic diagram showing the operation of a turbofan engine

Eventually, the British government presented Frank Whittle with £100,000 for his work in designing the first jet engines, and he was given a knighthood. He suffered several nervous breakdowns and then left the country, a disillusioned man, and went to live in the USA, where he died in early 1997.

Speed Through the Air: the Age of the Jet

Breaking the Sound Barrier

As an aircraft approaches the speed of sound, the sound waves it creates spread out in front, as they are travelling faster than the aircraft. But when sonic speed is reached, the waves cannot get out of the way, and become compressed. Shock waves then form around the aircraft, which can cause it to become unsteady, and buffeting can occur. The shock waves can be heard on the ground as 'sonic boom'. These can endanger the stability and

⊙ *A McDonnell Douglas F/A-18 Hornet breaking the sound barrier*

structure of the airframe, and control can be lost. To combat these problems, supersonic aircraft need swept wings to reduce what is known as 'wave drag', and to make the transition through the sound barrier smoother. If the speed of the aircraft reaches Mach 5, it is said to be 'hypersonic'.

Such high speeds for man-carrying 'aircraft' were first achieved by the series of experimental rocket-powered machines. These included:

Bell X-1: Single-seat research
Power plant: 1 x Reaction Motors RMI LR-8-RM-5 rocket, 6,000 lbf
Range: 4 mins 45 secs
Maximum speed: 1,450 mph (2,333 kmph)
Ceiling: 90,000 ft (27,450 m). The X-1 made twenty-six flights between September 1946 and June 1947. On 14 October 1947 Chuck Yeager broke the sound

The experimental aircraft Bell X-1 in flight

Speed Through the Air: the Age of the Jet

Chuck Yeager next to the Bell X-1 HASH 1 Glamorous Glennis

barrier for the first time while flying X-1-1 launched from the belly of B-29 bomber.

Bell X-2: Single-seat research
Power plant: 1 x Curtiss-Wright XLR25 rocket engine, 15,000 lbf at sea level
Wingspan: 32 ft 3 in (9.8 m)
Maximum speed: Mach 3.196 (2,094 mph, 3,370 kmph)
Ceiling: 126,200 ft (38,500 m)
Max takeoff weight: 24,910 lb.

North American X-15: air-launched from B-52 bomber
Flights made between 1959 and 1968
Power plant: XLR-99, anhydrous ammonia and liquid oxygen, reached Mach 6.7 on 3 October 1967.
Set altitude record of 354,200 ft (67 miles, 108 km) on 22 August 1963.

> A Mach number is defined as the speed of sound at sea level and a temperature of 15°C (59°F): 1,225 kmph (761 mph). As these parameters change, the speed of sound will change.

Once the problems associated with supersonic travel were solved, the knowledge gained was used to design practical warplanes capable of such speeds.

These are the fastest aircraft in the world:

Typhoon Eurofighter
Aircraft type: single-seat fighter
Wingspan: 35 ft 11 in (11 m)
Power plant: 2 x Eurojet EJ200 afterburning turbofan engines; dry thrust 60 kN (13,500 lbf) each; thrust with afterburner 90 kN (22,000 lbf) each
Maximum speed: at altitude, Mach 2 (1,320 mph); at sea level, Mach 1.2
Ceiling: 65,000 ft (20,000 m).

> An afterburner, or reheat, is an additional component added to some jet engines, primarily those on military supersonic aircraft. Its purpose is to provide a temporary increase in thrust, both for supersonic flight and for takeoff, as supersonic aircraft designs require a high take-off speed.

F-22A Raptor
Aircraft type: single-seat fighter
Wingspan: 44 ft 6 in (13.6 m)
Power plant: 2 x Pratt & Whitney F119-PW-100 turbofan engines with afterburners and two-dimensional thrust vectoring nozzles.

Thrust: 35,000 lb (16,000 kg each engine)
Maximum speed: Mach 2 with supercruise capability
Ceiling: above 50,000 ft (15,250 m).

Sukhoi Su-37
Aircraft type: single-seat fighter
Wingspan: 48 ft 3 in (15 m)
Power plant: 2 x Lyulka AL-37FU turbofans; 145 kN (32,000 lbf) each
Maximum speed: Mach 2.5 (1,550 mph, 2,500 kph) at high altitude
Ceiling: 59,100 ft (18,500 m).

Tupolev Tu-160 Blackjack
Aircraft type: world's largest operational bomber
Wingspan: 116.8 ft (36 m) swept, 82.7 ft (25 m) spread
Power plant: 4 x SSPE Trud NK-321 turbofans, 4 x 137 kN/ 245 kN
Maximum speed: 1,367 mph (2,200 kph)
Ceiling: 9.6 miles (15.4 km).

Lockheed SR-71 Blackbird
Aircraft type: Long-range strategic reconnaissance
Wingspan: 55 ft 7 in (15 m)
Power plant: 2 x Pratt & Whitney J58-1 continuous-bleed afterburning turbojets, 32,500 lbf (145 kN) each
Maximum speed: Mach 3.2 plus (2,200 mph plus, 3,540 kph plus) at 80,000 ft (24,500 m)
Ceiling: 85,000 ft (26,000 m).

The First Commercial Jet Services

When large propeller passenger aircraft, such as the Boeing Stratocruiser, were coming into service, the British company De Havilland produced an exceptional aircraft: it was a jet-powered airliner, the Comet. Here was an aeroplane designed to travel at twice the height of other airliners and at the speed of a Second World War fighter.

The Comet first flew in 1949 and, in May 1952, went into service with BOAC. The first trans-Atlantic service followed in 1958, but two serious crashes delayed development. However, Comets were operated by twenty-four airlines and they remained in passenger service until 1981. During this time Boeing began its domination of long distance commercial flying with their 707 jetliner, which itself

Qantas Boeing 747 landing at Heathrow airport, London

entered trans-Atlantic service only three weeks after the Comet. Ultimately, in 1970, the Boeing 747 'Jumbo Jet' (2½ times larger than the Boeing 707) went into service and the huge manufacturing power of the American company made these aircraft the most successful passenger carriers so far.

Concorde

The first supersonic airliner to enter service was the Aérospatiale–BAC Concorde. Although conceived in Great Britain in the early 1960s to transport one hundred passengers at supersonic speeds across the Atlantic, it became a joint Anglo-French project, largely because of the huge development costs involved in the project.

> When the Concorde project was started, supersonic flying had been proved a practical proposition. Although the ability to break the sound barrier was limited to military jets, there were foresighted engineers who could see the advantages of passengers being transported in even faster times than the jetliners that were coming into service. They knew it was going to be a long haul and very expensive but very worth while.

The design for Concorde was a delta-wing configuration powered by four Roll-Royce/SNECMA Olympus 593 Mk 610 afterburning jet engines; these were based on the units that powered the Vulcan bomber. The design also provided for the first airliner fitted with an electronic 'fly by wire' flight control system. It was intended to fly at 60,000 ft (18,300 m)

Concorde's last-ever landing at Filton Airfield, Bristol, 2003

at a cruising speed of Mach 2.02 (1,330 mph, or 2,140 kmph) at its cruising height. This resulted in London–New York flights taking just 3 hrs 30 mins.

The first test of a prototype, built by Aérospatiale in France, took place at Toulouse on 2 March 1969, with the first supersonic flight taking place on 1 October that year. The first British-built Concorde flew on 9 April 1969, with the first commercial flights commencing in 1976 and continuing until 2003.

Among the problems that had to be solved in the design of Concorde was that of controlling the temperature of the airframe at speeds of up to Mach 2.2. During Concorde's climb to its cruising altitude its surface cooled down but, when it went supersonic, the surface heated up and, in so doing, the fuselage increased in length by almost one foot. As the airframe was largely made of aluminium, the highest temperature that could be sustained was 127°C (260.6°F); this in turn determined the maximum speed of the aircraft. The airframe temperature was controlled by passing cool fuel around the wings and parts of the fuselage when required.

Although orders were made for seventy aircraft only two prototypes, two pre-production and sixteen actual Concordes were built. Of these, fourteen were flown commercially, by Air France and British Airways. One of the French Concordes crashed in Paris on 25 July 2000. Eight were still in service in April 2003 when operations ended, partly as a result of the effects on the world economy arising from the 9/11 attacks. In addition, with the aircraft approaching thirty years of age, a large investment programme to update the fleet became uneconomical.

Concorde was not the only supersonic passenger aircraft to be flown; the Soviet Union built the Tupolev Tu-144, known as 'Concordski' for its close resemblance to Concorde. Following two crashes, including one at the Paris Air Show in 1973, it was not considered to be a commercial proposition. Two American companies expressed interest in a supersonic transport, including Boeing with its SST, and one from the North American company, but they did not proceed beyond the preliminary design stages.

Air Speed Records

These are the speed records for piloted aircraft from 1903 to the present day:

Year	Pilot	Aircraft	Speed mph
1903	Wilbur Wright	Wright Flyer	9.80
1905	Wilbur Wright	Wright Flyer III	37.85
1908	Henri Farman	Voisin	40.26
1909	Louis Blériot	Blériot XII	47.82
1910	Alfred Leblanc	Blériot XI	68.20
1911	Edouard Nieuport	Nieuport Nie-2	82.73
1912	Jules Vedrines	Monocoque Deperdussin	108.2
1913	Maurice Prevost	Monocoque Deperdussin	126.7
1914	Norman Spratt	RAF SE4	134.5
1918	Roland Rohlfs	Curtiss Wasp	163.1
1919	Joseph Sadi-Lecointe	Nieuport-Delage	191.1
1920	Joseph Sadi-Lecointe	Nieuport-Delage	194.5
1921	Joseph Sadi-Lecointe	Nieuport-Delage	205.2
1922	Billy Mitchell	Curtiss R-6	224.3
1923	Alford J. Williams	Curtiss R2C-1	267.2
1924	Florentin Bonnet	Bernard Ferbois V2	278.5
1927	Mario de Bernardi	Macchi M.52	297.8
1928	Mario de Bernardi	Macchi M.52 bis	318.6
1929	Giuseppe Motta	Macchi M.67	362.0
1931	George Stainforth	Supermarine S.6B	407.5
1933	Francesco Agello	Macchi MC.72	424.0
1934	Francesco Agello	Macchi MC.72	440.6
1939	Fritz Wendel	Messerschmitt Bf 209	469.22
1941	Heini Dittmar	Messerschmitt Me 163	623.65
1944	Heinz Herlitzius	Messerschmitt Me 262	624.0
1944	Heini Dittmar	Messerschmitt Me 163	702.0
1945	H. J. Wilson	Gloster Meteor F Mk 4	606.4
1946	Edward Donaldson	Gloster Meteor F Mk 4	615.78

continues opposite

Year	Pilot	Aircraft	Speed mph
1947	Albert Boyd	Lockheed P-80R Shooting Star	623.74
1947	Chuck Yeager	Bell X-1	670.0
1948	Richard L. Johnson	North American F-86A-3 Sabre	670.84
1953	Neville Duke	Hawker Hunter F Mk 3	727.6
1953	Mike Lithgow	Supermarine Swift F4	735.71
1955	Horace A. Hanes	F-100C Super Sabre	822.1
1956	Peter Twiss	Fairey Delta 2	1,132.0
1959	Georgii Mosolov	Prototype MiG-21	1,484.0
1961	Robert L. Stephens and Daniel Andre	Modified F-4 Phantom	1,606.3
1976	Eldon W. Joersz	SR-71 Blackbird	2,194.0

The 1976 speed record remains unbroken at the time of writing.

Measuring Speed

The first ways of measuring speed over land were simple. Hand signals and a stopwatch were used to provide the times of a vehicle entering and leaving a measured distance. A simple set of tables would convert the time difference into an average speed over the distance. This method suffered from inaccuracies of the human hand and eye. When determining speeds for record purposes, the time for covering, for example, a measured mile is measured twice in opposite directions. This is intended to cancel out any effect that the wind may have.

Instruments that indicate speed to a driver or pilot while a car, boat or aircraft is moving have been around for more than a hundred years. In a car the speedometer is a device, now usually fully electronic, connected to one of the wheels or to the engine drive shaft. The speedometer measures and then converts speed into a visible readout in miles per hour, or kilometres per hour. On a ship or aircraft, speed is measured by a tube, known as a 'pitot head'. One end of the tube is open and senses the pressure of the air, or water flow, while an opening at 90° to the direction of motion measures the static air or water pressure. The difference in pressures is converted into the speed of the boat or aircraft.

Some relative speeds
Snail, common: 0.002 mph (0.004 kmph)
Man, brisk walk: 4 mph (6 kmph)
Olympic sprinter, av. speed over 100 metres: 22 mph (36 kmph)
Racehorse, at gallop: 25–30 mph (40–48 kmph). The world record for a horse galloping over a short distance is 55 mph (89 kmph)
Cheetah, at top speed: 70 mph (113 kmph)
Peregrine falcon, in full flight: 200 mph (322 kmph)
Boeing 747-8, cruising speed: 650 mph, Mach 0.85 (1,050 kmph)
Sound (dry air, sea-level pressure, 20° C): Mach 1, 768 mph (1,235 kmph)
Escape velocity, on Earth: 18 miles/sec (11.2 km/sec)
Space Shuttle, speed on re-entry: 17,500 mph (28,000 kmph)
Average orbital speed of planet Earth: 66,623 mph (107,218 kmph)
Light (in vacuum): 914,367,000 feet/sec (299,792,458 metres/sec).

Variations on the Theme of Flight

Rotating wing machines
In the fifteenth century Leonardo da Vinci designed a machine that would rise vertically from the ground –

what we would now call a helicopter – with a man-operated helical screw to provide the lift. Being made of wood, it would never have achieved flight, but the concept was one that was followed in later years.

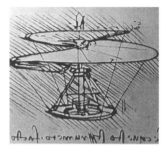
⊙ Leonardo's drawing of a helicopter

Many attempts to build helicopters were made in the late nineteenth and early twentieth centuries, but it was not until 1907 that one actually lifted itself and its pilot off the ground.

> In November 1907 a helicopter with a pair of counter-rotating wings, designed by Paul Cornu, lifted off in Lisieux, France, but further development stopped for lack of funds. Other similar flights were made in the following years but they too were never proceeded with.

Juan de la Cierva first flew a rotating wing aircraft, an autogyro, in 1923 in Madrid, Spain. It consisted of a conventional fuselage and tail with a forward mounted engine. Above the pilot was a set of slim wings that were free to rotate. As the plane moved forward these wings began to rotate until after forty yards they produced enough lift to take off.

The first true helicopter was designed and built in America in 1939, by Igor Sikorsky, with a combination of one main rotor, providing the lift and forward movement, and one tail rotor to compensate for the torque that is produced by the main rotor. The tail rotor also controlled the helicopter along the vertical axis during hover flight. It is this basic configuration that has become the template for the majority of helicopters in use today.

The first successful helicopter with a pair of powered rotating wings, each one cancelling out the torque of the other, was the Focke-Wulf Fw 61 that in 1938 was dramatically demonstrated inside the Deutschlandhalle in Berlin and flown by the famed Hanna Reitsch. This particular helicopter was one of the few with a pair of rotors outboard of the fuselage. A number of twin tandem rotor helicopters have been built, the most successful of which is the twin-engined Boeing CH-47 Chinook, that is mainly used as a military heavy-lift helicopter. Known as the 'flying banana', it has a top speed of 196 mph (315 kmph).

> In the operation of helicopters and autogyros, 'autorotation' is the name given to the generation of lift by the main rotor, even when no power is being provided to the rotor by an engine. Following an engine failure, a helicopter may be able to slow its descent before landing, and land in a controlled manner, using autorotation.

When is a helicopter not a helicopter? When it's the V-22 Osprey. This unique machine can take off like a helicopter and fly like a conventional aeroplane, and it does this by having tilting rotors. Its pair of engines, outboard of the fuselage, are vertical for a helicopter-like take-off. Once in the air these swing down through 90° to allow forward flight. A prototype V-22 first flew in 1988, and the aircraft went into operation in 2005. During its life it has been bedevilled with controversy over its safety record. This is because if it is in helicopter mode at a height of under 1,600 ft (490 m) and suffers complete engine failure it cannot do what a helicopter can do under these circumstances: that is, it cannot autorotate and will inevitably crash. At a higher level the tilt rotors can be rotated to the horizontal and the aircraft can glide to earth.

The V-22 taking off from Alliance Airport, Fort Worth, Texas, in 2006

The only true vertical takeoff and landing aircraft that can revert easily to horizontal flight, forwards or backwards, is the British Hawker Siddeley Harrier (AV-8A in the USA) fighter. This uses a single vectored thrust turbojet engine whose jet outlets can be rotated through 90° in order to change the direction of flight. It first flew in 1967 and is still in active use by a number of air forces.

Flying cars

Comic books in the 1930s and '40s featured the ultimate in personal transport – the flying car. A personal plane is fine, except that you start and finish on an airstrip, and to get to the airstrip you need a car, and another one at the end of your flight. Combine the two machines and you can go where you will. A number of attempts were made to build such a machine but it was not until 1937 that the Waterman Aeromobile took to the road and to the air. Unfortunately, as with other early flying cars, the machine was clumsy, needing wings that had to be towed on a trailer behind the car. The wings would then be bolted on to the car body, and the engine declutched from the wheels and connected to the propeller drive. Modern technology now provides a better way of compacting this type of machine.

- The Terrafugia Transition, now being developed in America, is a straightforward car with wings. Its 27 ft (8.25 m) wings fold up close to the body, and the engine drives a rear-mounted propeller. It has a range of 500 miles (804 km), cruising at 115 mph (185 kmph). Soon to be on sale.

- Also under development is the PAL-V, a three-wheeled gyrocopter that can take off and land vertically, with a flying speed of 125 mph (201 kmph) and a ceiling of 4,000 ft (1,220 km).

- The Moller Skycar M400 is at the concept stage at present, but a prototype has been demonstrated.

It is powered by eight Wankel rotary engines that rotate from the horizontal, for road use, to the vertical (as with the V-22 Osprey) in order to take off and land. When in the air they revert to the horizontal for flight, cruising at 29,000 ft (8,845 km) at 300 mph (483 kmph) and with a maximum speed of 380 mph (611 kmph). An emergency parachute is fitted in case of an airborne failure.

- The Skycar: Richard Fleury, a British engineer, has invented a fan-powered flying car. To prove that the Skycar works, in 2009 he travelled from the United Kingdom to Timbuktu in the West African state of Mali, partly on land and partly in the air. The Skycar is equipped with a parafoil (a non-rigid textile airfoil wing) stowed in the boot and deployed from the rear of the car before takeoff when the car's transmission can be moved from road mode to flight mode. When in flight mode the engine powers a rear-mounted fan. The thrust from the fan provides lift for the parafoil and, at a speed of 35 mph, it takes off in 200 yards. Pedals in the 'cockpit' steer when in the air by changing the shape of the wing. In the event of the wing being damaged there is an emergency parachute that allows the Skycar to drift down to earth.

Engine: 1000 cc, four cylinders. Power: 140 bhp
Range: 180 miles (290 km) in flight, 250 miles (402 km) on road
Cruising ceiling: 2,000–3,000 ft (610–915 m)
Ceiling: 15,000 ft (4,575m). Max. speed: 80 mph (129 kmph) in flight, 110 mph (177 kmph) on road.

› Chapter 5

The Future

Rockets

For many years after the First World War experiments were carried out in Germany to see if it would be possible to power vehicles by rockets. These experiments had a certain amount of success with rocket-propelled cars, but it was the outbreak of the Second World War that gave rocket research the impetus that seriously boosted research. In the early 1940s the Germans produced a successful rocket-powered aeroplane – the Messerschmitt Me 163, a single-seat attack fighter designed to down American B-17 Flying Fortress bombers.

There were, however, two other developments that were intended to play a decisive role in the German war effort: one was the V1 flying bomb; the other was the V2 supersonic missile. The V1, known as the 'buzz bomb', or 'doodlebug', was the Fieseler Fi 103, a small, pilotless, flying bomb that was powered by a 'pulsejet' engine and fired from a ramp towards its objective. When its fuel ran out it fell to earth and exploded on impact.

The Future

> A pulsejet was a form of rocket that was a long tube whose front was covered by a series of shutters. An air-gasoline mixture was injected into the tube ignited by a spark plug. The explosion forced the shutters shut and propelled the missile forward. Then the air pressure opened the shutters and the firing restarted, continuing a cycle of explosions.

The V2 was a true rocket, officially known as an A4 and designed by Werner von Braun. It weighed approximately 13.6 tons and contained a ton of explosive in its warhead. The rocket motor, fuelled by a mixture of alcohol and liquid oxygen, produced a thrust of twenty-five tons to launch the missile, and quickly reached supersonic speeds.

V-2 Rocket in the Peenemünde Museum, Germany

After the war, Von Braun and some of his rocket team were taken to the United States. Ten years after entering the country, he became a naturalised US citizen. He worked on the American intercontinental ballistic missile (ICBM) program before joining NASA, where he served as director of NASA's Marshall Space Flight Center and as chief architect of the Saturn V launch vehicle that propelled the Apollo spacecraft to the Moon. He is generally regarded as the father of the United States space program.

The Messerschmitt Me 163B Komet rocket-propelled fighter

The lethal Messerschmitt Me 163B, introduced towards the end of the war, was the only rocket-powered combat aircraft of the Second World War. It had no undercarriage, was launched from a trolley and landed on a retractable skid. The operational version was powered by an engine fuelled by a mixture of hydrazine hydrate and methanol, designated C-Stoff, and T-Stoff, hydrogen peroxide. This highly dangerous fuel took it up to 40,000 ft (12,200 m) in three minutes, at a speed of

550 mph (900 kmph). It could then dive on to the enemy bombers, attack them and then, with the fuel exhausted, glide back to base.

Me 163B
Crew: 1
Length: 18 ft 8 in (5.7 m)
Wingspan: 30 ft 7 in (9.33 m)
Wing area: 200 ft^2 (186 m^2)
Maximum takeoff weight: 9,500 lb (4,310 kg)
Maximum speed: Mach 0.83, 596 mph (960 kmph)
Duration of flight: 8 mins
Range: 25 miles (40 km)
Service ceiling 39,700 ft (12,108 m)
Rate of climb: 525 ft/sec (60m/sec)
Armament: two 30 mm Rheinmetall Borsig MK 108 cannon, 60 rounds per gun.

There now comes a problem that researchers are attempting to solve. More and more effort is being put into finding faster means of travel – but why? People want to travel fast for one of two reasons: either they need to get from one place to another as quickly as possible, or they want to go as fast as possible just for the experience. Providing a service for people who need high-speed travel, as Concorde did, can be achieved but at a high cost. To travel to New York from London by Concorde cost the first class fare plus a further 30 per cent. High-speed trains, however, because there are more of them in use and many more people use them, cost little more, if anything, than ordinary fares. So there has to be a balance between the need for speed and the increase in comfort.

Much research is going on examining the possibilities of higher-speed air travel using newer means of propulsion. One projected type of engine is the ramjet, an air-breathing engine with no moving parts, that requires a forward speed of around Mach 3 to operate efficiently, and which can accelerate to speeds of at least Mach 5. It uses the aircraft's forward motion to force air into an opening that surrounds a pointed plug in the front of the fuel chamber. This compresses the air as it enters the fuel chamber, where it burns and creates thrust. The supersonic flow of air into the engine is decelerated at the inlet to subsonic speeds, and then reaccelerated through the nozzle to supersonic speeds to produce thrust.

The scramjet

A further development of the ramjet is the scramjet (supersonic combustion ramjet), which requires a supersonic airflow through the engine. This is thought to be the way into hypersonic travel. A very simple scramjet would look like two funnels attached by their small ends. For a vehicle that carries a scramjet engine to work, it must reach Mach 7 before it can function. At supersonic speed, air is forced into the first funnel, compressing and heating in the process. A fuel, such as hydrogen, is added at the narrow section, where the funnels join and compression is greatest, and burned. This heats the gas further, which then expands and

The Future 151

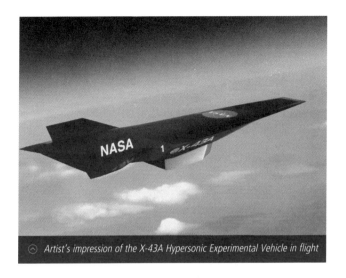

Artist's impression of the X-43A Hypersonic Experimental Vehicle in flight

exits through the second funnel, as in a rocket, thus producing thrust. Research suggests that the optimum speeds produced by a scramjet lie from Mach 12 to Mach 24. If a hypersonic plane powered by a scramjet went into commercial service, then Tokyo could be reached from New York in two, rather than eighteen, hours. The practicalities of such an aircraft have yet to be investigated.

In November 2004 the unmanned X-43A set a new speed record of 7,546 mph (12,140 kmph), or Mach 9.8. It was boosted by a modified Pegasus rocket, and launched from a Boeing B-52 at 43,166 ft (13,179 m). After a free flight, where the scramjet operated for about ten seconds, the craft made a planned crash into the Pacific Ocean, off the coast of southern California.

Space Travel

Serious space travel is still a long way off, although plans are under way for further landings on the Moon, and even to Mars. These flights will allow research into the possible establishment of permanent bases there. Such bases would allow the Moon, in particular, to act as a jumping-off ground for future space flights.

Moon landings

The surface-to-surface journey to and from the Moon is approximately 75,860 km (114,637 miles). The Apollo 11 mission, from takeoff to landing, lasted 107 hours. The first stage of the launch rocket accelerated Apollo 11 to a speed of

Astronaut 'Buzz' Aldrin walking on the surface of the Moon, 20 July 1969

6,000 mph (9,700 kmph) and the second stage, in order to reach orbital position 115 miles (185 km) above Earth, boosted the speed to 15,800 mph (30,000 kmph), before heading for the Moon. These speeds are only obtainable in the vacuum of space, where there is no friction over the surface of the vehicle and the power source is non-airbreathing.

Visiting Mars

The distance from Earth to Mars is a variable, but is approximately 100,000,000 miles (160,000,000 km), and a journey there would take about five and a half months. The crew of a visiting spaceship would then have to wait some twenty months before returning to Earth. This is because the relative positions of Earth and Mars needed for a return journey occur in a cycle. With the return journey lasting a further five and a half months, the whole operation would last almost three years. The average speed of the journey would be around 25,000 mph (40,000 kmph). Only the g forces on landing and takeoff would cause discomfort, but there are other physical problems that would have to be overcome.

A lengthy journey in space will have serious effects on the human body. Bones will deteriorate in zero gravity, muscles will atrophy, and there could be a decrease in the body's production of red blood cells. Those apart, there would also be psychological problems associated with a group of perhaps five persons being confined in a capsule for six months at a time.

In addition to these problems is the question of whether or not time speeds up or slows down the aging process when travelling a long distance in space. It is thought that while travelling through space an astronaut's chromosomes are exposed to penetrating cosmic rays, which can damage the body's telomeres.

> A telomere is a region of repetitive DNA capping the ends of chromosomes, which protects them from destruction. They shorten each time a cell divides. Their loss is believed to be linked to aging. When they become too short, the cell dies.

The Space Shuttle

The US Space Shuttle programme began in the 1970s, and is scheduled to finish in 2010. Five shuttles were built: *Atlantis*, *Challenger*, *Columbia*, *Discovery* and *Endeavour*. They have completed a grand total of 1,180 days in space, and have flown 484,317,674 miles (779,267,137 km) while in orbit. Tragically, both *Challenger* and *Columbia* were lost in accidents.

After the termination of the Space Shuttle programme, the Orion programme will begin. The Orion vehicle will be capable of ferrying astronauts and scientists to the International Space Station, currently in orbit 220 miles (350 km) above Earth, and travelling at 17,227 mph (27,724 kmph). However, the main purpose of this spacecraft will be to carry astronauts out of our orbit and on to other worlds. It will be the first spacecraft of its kind since the Apollo Command Module. Orion will be configured in a

The Future 155

Atlantis *breaking the sound barrier on launching in 2000*

similar way, but will be three times the size and hold up to six crew members. As before, the Orion module is designed to return to Earth, landing by parachute on water. The first flight is scheduled for 2014.

Commercial space travel

However, space travel for business and pleasure will be decades away. Nevertheless, there are projects in place that will provide, for those who can afford it, the opportunity to experience travel into the edge of space. In the forefront of one project is Richard Branson, the billionaire British businessman and adventurer. His new company, Virgin Galactic, is providing a chance for people to have an experience of space travel in the very near future.

> **Virgin Galactic**
> So far over 65,000 people have applied for the first batch of one hundred tickets, paying a deposit of US $200,000 each. Before taking a flight, they will follow a three-day aptitude course and pass a high-g centrifuge test.
>
> A total of 44,000 people have registered for subsequent flights. Two mother ships are to be built, together with five spaceships; weekly launches are planned.

Branson, in cooperation with aircraft designer Burt Rutan, unveiled in 2008 an aircraft that is designed to serve as the mothership for the world's first commercial suborbital flight. The mothership, called *White Knight Two* (WK2), will launch the first space flight. It is a conventional aircraft, with two fuselages joined by a wing, and resembling a four-engined flying catamaran. Beneath the wing that joins its two halves is slung an eight-person rocket ship, known as *Space Ship Two*. In 2010 the whole machine will fly to a height of 48,000 ft (15,000 m) at which height *Space Ship Two* will be launched. Under its own power the rocket ship will climb up to a height of 360,000 ft (110,000 m) at a speed of 2,600 mph (4,200 kmph) and for four minutes it and its passengers will cruise at the edge of space and experience weightlessness. It will then glide back to Earth. The entire flight will take two and a half hours.

> The precursor of *Space Ship Two* was Burt Rutan's *Space Ship One*, which made its first test flight in June 2004.

The Future of Travel: Faster, Further, Smarter?

So where is all this leading us? We must first of all consider what people really want. The growing need for mass travel is already with us, and research tells us that high-speed air travel is made less attractive by the time taken getting to and from airports, and by the congestion that results from the stacking of flights taking off and landing at busy terminals. One possible solution could be for passengers to take point-to-point-on-demand services between small airports. Given forecasts that air traffic will treble by 2025, there will be a need for a large number of small local airports. In the USA, 98 per cent of the population already live within a thirty-minute drive of one of the 5,400 small local airports. This might just be the catalyst that would make use of these airports more attractive, particularly if suitable vertical takeoff and landing (VTOL) aircraft could be used, where no long runways exist. By this means, point-to-point-on-demand 'air taxis' might become a very practical proposition.

At present we are seeing the creation of very large passenger aircraft such as the European Airbus A-380. Here we have a double-deck, wide-bodied, Superjumbo aircraft that can carry more than five hundred passengers with 50 per cent more floor space than a Boeing 747-400. By carrying only economy class passengers, an A-380 can seat more than eight hundred people. Its size creates problems, as there is a limited number of airports that can cope with it. However, use of this type of large aircraft does mean that the concept of hub-and-spoke airports becomes an option.

The solution offered by Boeing is the 787 Dreamliner, a fuel-efficient, mid-sized, wide-body, twin-engined jet airliner that will finally come into service in 2010. The three versions of the aircraft will carry between 210 and 330 passengers. Its design will enable it to use airports both large and small.

Model of the Boeing 787 Dreamliner

Of course, air travel is not the only way to get from place to place. If people want to see the world they will still have railway trains and ocean liners to transport them in a more leisurely manner. High-speed train services are in operation, or being planned, in almost every European country as well as in Argentina, Brazil, China, Mexico, Saudi Arabia, South Korea and the USA. In Spain a high-speed rail link is being built to Malaga airport.

A report from LET (Laboratoire d'Economie des Transports), the French multidisciplinary transport research unit, states that in the last few decades many European airports have become interconnected with European high-speed railway networks. However, this interconnection can really only be practical for the biggest airports.

In America the operation of high-speed trains has been slow to take off. But it is said that if high speed rail is implemented correctly, as in many European countries, with rail lines running right into airport terminals, transfer between plane and train will be seamless and render the need for flights of less than 500 miles unnecessary in most cases.

In Europe the move is towards high-speed train links between major cities, and already the need for more of these is clear as more and more passengers are taking trains rather than short distance flights.

Travel in a more leisurely style is now offered by an increasing number of cruise liners. These give the appearance of huge, floating apartment blocks, bearing no resemblance to the stately Atlantic liners of the 1930s. Again, exploring the world by sea is becoming an attractive way for people to see exotic parts of the world that they would never have thought of visiting. Flying is fine if you want to cover a long distance quickly, but cruising, even on a high-speed train through the Italian or Spanish countryside, does have its attractions. It's going to be a pick-and-mix world of travel before long.

Or Will it be Greener?

Despite all the work going on in attempting to travel faster and with less stress, there is the additional need to balance our use of the earth's resources needed to power our travel. Fossil fuels will disappear before very long if we keep travelling the way we do, and so alternatives will have to be found to find to keep the engines going. Added to which, there is no way we can continue to allow the pollution that results from our continued burning of fossil fuels. We therefore have to consider ways of powering our cars, boats and aircraft in a sustainable way that does not produce excesses of, and preferably does not produce any, carbon dioxide (CO_2).

> Fossil fuel is fuel, such as coal, oil and gas, that has been created beneath the Earth's surface over millions of years.

Diesel
Diesel engines of the latest designs often have lower CO_2 emissions and better fuel economy than petrol engines, but they emit other pollutants that affect air quality. When and where we use a diesel engine is important in tackling air quality issues. Driving at a constant speed along motorways in a diesel-powered vehicle is less damaging to the atmosphere than stop-start driving in a city.

Liquid propane gas
LPG provides good fuel economy and lower CO_2 emissions. Propane combustion is much cleaner than gasoline combustion, though not as clean as natural gas combustion, creating organic exhausts besides carbon dioxide and water vapour during typical combustion.

Electric cars

Running on rechargeable batteries, electric cars are certainly quiet and make no emissions. However, their batteries have to be charged from an electricity supply that itself comes from a fuel-burning source – unless nuclear power stations become more popular. The Tesla Roadster is an electric sports car that is capable of reaching 130 mph (209 kmph); but the faster it travels, the shorter the battery life. Also, recharging the Tesla's batteries takes several hours.

Tesla Roadster

Engine: 3735 V AC motor powered by lithium-ion batteries

Power: 248 bhp. Top speed: 130 mph (210 kmph)

Acceleration: 0–60 mph (97 kmph) in 3.9 secs

Cost of a 'top-up' is one-tenth of the equivalent in petrol

Fuel hybrid vehicles

These are equipped with a petrol engine that provides the main motive power, with an electric motor assisting when needed. The efficiency of the technology means they emit lower CO_2 emissions than ordinary petrol or diesel vehicles.

In a hybrid vehicle the electric motor is used at low speeds, with the petrol engine cutting in at above-average urban speeds. The system is regenerative so that when the car is coasting or semi-coasting on a feathered throttle, the engine generates electricity which is stored in a battery pack.

Biofuels

The 'Renewable Transport Fuel Obligation' (RTFO) has placed an obligation on fuel suppliers, from April 2008, to ensure that a percentage of their total sales are of biofuels. In the UK, it is a requirement that 5 percent of all fuel sold on petrol station forecourts must come from a renewable source by 2010.

Biodiesel

Biodiesel uses vegetable oil as a substitute for conventional diesel, and can reduce CO_2 emissions by around a half. Most diesel-engined cars will operate well on a 5 percent mix (B5) but manufacturers' warranties may be invalidated.

Bioethanol

A liquid biofuel made from fermented, distilled starch plants such as corn or sugar, bioethanol is widely used in Brazil. It can be used on its own or blended with petrol. Ethanol is of a higher octane value that petrol and has been used as a fuel for racing cars.

> The octane rating of petrol indicates how much the fuel can be compressed before it spontaneously ignites. The higher the octane rating, the higher the compression of the fuel, and the greater the power generated.

Biogas

Biogas is a renewable alternative fuel, produced by breaking down organic matter through a process of microbiological activity. Rotting municipal waste, food waste or sewage (both human and animal), is turned into gas containing methane and carbon dioxide by means of 'anaerobic conversion' in a digester.

Fuel cells

These are power sources that produce electricity in a type of battery. Unlike a conventional battery, which reproduces the electricity that has been stored within it, a fuel cell produces electricity from the chemical reaction between a flow of fuel, such as hydrogen, and an oxidant, usually oxygen, which react in the presence of an electrolyte. Fuel cells can operate continuously as long as the necessary flows are maintained. The electrodes within a fuel cell, unlike those in a battery, act purely as catalysts and remain stable during use. The state of California is now experimenting with hydrogen fuel-celled buses.

Solar power

Solar power is generated by power cells, commonly used in solar panelling, that convert the sun's energy into electricity. They are made up of semiconductors, usually of silicon, that absorb the light. The sunlight's energy then frees electrons in the semiconductors, creating a flow of electrons that generates the electricity that charges a battery, or powers a motor that drives wheels.

Semiconductors are usually made from silicon, a solid material that can both conduct electricity and act as an insulator.

At the time of writing a Swiss adventurer is preparing to build a solar-powered aeroplane that would be able to store enough electricity to enable it to remain airborne at night, reach a maximum height of 28,000 ft (8,500 m) and fly at a speed of 44 mph (70 kmph) around the world. Students from Cambridge University have built a solar-powered car capable of 60 mph (97 kmph).

Hydrogen

Hydrogen can be used as a fuel to power an internal combustion engine instead of carbon-based fossil fuels. It is obtained by the electrolysis of water. (In electrolysis an electric current is passed through water. Hydrogen and oxygen can then be collected from the electrodes.) A hydrogen internal combustion engine is estimated to be 38 per cent efficient, 8 per cent higher than a petroleum-powered engine. The only substance that emerges from the exhaust pipe is water.

Only when new fuels are perfected can we take a step towards changing dramatically the face of mass transport. The world might then become even smaller than it is today but, in the process, it will become a much better place in which to live.

Further Reading

Foxworth, Thomas G. *The Speed Seekers*. Haynes Publishing Group, 1989.

Guinness Word Records. Guinness Plc, published annually.

Sells, John A. *Stagecoaches across the American West. 1890–1920*. Hancock House Publishers.

Mackworth-Praed, Ben (researched and edited). *Aviation – The Pioneer Years*. London: Studio Editions, 1992.

Humble, Richard (ed.). *Naval Warfare – An Illustrated History*. Orbis Publishing, 1983.

Quinn, Tom. *Wings over the World. London:* Aurum Press, 2003.

Wilson, Hugh. *Encyclopaedia of the Motorcycle*. Dorling Kindersley, 1995.

Garnett, A.F. *Steel Wheels (The Evolution of the Railways & How They Excited Engineers, Architects, Artists & Writers)*. Cannwood Press, 2005.

Standage, Tom (ed.) *The Future of Technology*. Profile Books, 2005.

Nye, Douglas. *Carl Benz and the Motor Car (Pioneers of Science & Discovery)*. Priory Press, 1973.

Clifton, Paul. *The Fastest Men on Earth*. Herbert Jenkins, 1964.

Rendall, Ivan. *The Chequered Flag – 100 years of motor racing*. Weidenfeld and Nicolson, 1993.

Mountfield, David. *Stage and Mail Coaches*. Shire Publications, 2003.

Lucsko, David N. T*he Business of Speed*. Johns Hopkins University Press, 2008.

Index

Figures in italics indicate captions; those in bold indicate tables.

aeolipile 17, *17*
aerofoil 76, 77
afterburner (reheat) 132
air speed records **138–9**
airships 81–5, *82*, *84*, 100
al-Khowarizmi, Mohammed ibnMusa 15–16
Alcock and Brown (John Alcock and Arthur Whitten Brown) 99, 100
Alexeev, Rostislav 75
Arnold, General 'Hap' 123–4
Atlantic crossing 64–9, 99
Australia, travel to 70–71, 114–17
autogyro 141, 142
autorotation 142

balloons 78–81, *79*, 102
Beaverbrook, Lord 123–4
Bell Aircraft 124
Bell X-1/X-2 aircraft 130–31, *130*
Benz, Karl 32, 33
Benz Viktoria 32, *32*
Bettmann, Siegfried 49
bicycles 46–8, *47*, *48*, **48**, 49
biodiesel 162
bioethanol 162
biofuels 162
biogas 162–3
Blanchard, Pierre 79–80
Blériot, Louis 95, *95*
Bluebird 37, *37*, 73
Boeing 707 jetliner 134–5
Boeing 747 'Jumbo Jet' 135
Boeing 787 Dreamliner 158, *158*
Branson, Sir Richard 68, 102, 155–6
British Overseas Airways Corporation (BOAC) 116, 117

Brunel, Isambard Kingdom 23–4, *23*, 29
Brunel, Marc Isambard 29
Butterfield, John 14

Cairo to Cape flight 103–5
Campbell, Donald 45, *45*, 73–4
Campbell, Sir Malcolm 37, *37*, 73
canoes 52
carracks 60
cars
  electric 161
  flying 144–5
  fuel hybrid vehicles 161
  see also internal combustion engine
catamarans 75
Catch-me-who-can locomotive 20, 25
Channel Tunnel 28–30, 75
Chichester, Sir Francis 71
Chinook helicopter (Boeing CH-47) 142
Chitty Bang Bang 36, *37*
chronometer 56
Churchill, Sir Winston 102, 123–4
civilian flying, the first (1918–52) 96–8, *96*
clippers 57–9, *57*
Cobham, Sir Alan 104–5
Cockerell, Christopher 74
Columbus, Christopher 64–5
Comet aircraft 134, 135
commercial jet services, first 134–5, *134*
Concorde 135–7, *136*, 149
Cook, Captain James 70
Cornu, Paul 141
Cripps, Sir Stafford 127
Cutty Sark 58

Daimler, Gottlieb 49
dandy horse 46–7, *47*
de Dion, Jules 34

De Havilland light bombers, as civilian aircraft 96, 97
de la Cierva, Juan 141
Diesel, Dr Rudolf 27
diesel engine 27, 160
dirigibles 81
dugouts 52

Earhart, Amelia 109–12, *111*
ekranoplans 75
English Channel, crossing 95, 96
European Airbus A–380 157–8
Eurostar shuttle service 30, *30*
Eurotunnel shuttle service 30

F-22A Raptor 132–3
First World War aircraft 87–9, **87**, *89*
Fleury, Richard 145
'fly by wire' control system 135
Flyer 85, *86*
flying boats 101, 102, 112, 113, 115
flying eastwards (1918–39) 103–9
Flying Scotsman service 27–8
fossil fuels 160
fuel cells 163
funerary (burial) boats 53

galleons 60
galleys 59–60, *59*
General Electric 124
Giffard, Henri 81
Gloster Aircraft Company 120
Gloster Meteor 123, 127
Gotthard Tunnel 29
Graf Zeppelin airship 83
green fuels 159–63
Gresley, Sir Nigel 27–8
ground effect vehicles 74–5
Guinness, Kenelm Lee 37

Index

Harrison, John 56
Hawker Siddeley Harrier (AV-8A) fighter 143
Heinkel, Ernst 122
Heinkel He 162 aircraft 126, 127
Heinkel He 178 122–3
helicopters 140–42, *141*
Henshaw, Alex 107–8
Hero of Alexandria 17, *17*
Hindenburg airship 83–5, *84*
Hitler, Adolf 126
Hitler Youth 126–7
hobbyhorse (draisine) 46, *47*
hovercraft 74–5
Huskisson, William, MP 22
hydrofoils 75
hydrogen 164
hypersonic aircraft 130, 151, *151*

Imperial Airways 98, *98*, 104, 114, 115, 116
Imperial German Army Air Service 87, 88
Industrial Revolution 15, 17, 20
internal combustion engine 32–46
   Brooklands 36–8, *37*
   Grand Prix races 40–44, **40**, *41*, **41**, **42**
   the horseless carriage 32–3, *32*
   Indianapolis 38–9
   land speed record 36, 37, 44–6, *45*
   Le Mans 39–40
   racing improves the breed 33–46
ironclads 62
Isle of Man TT road races 50–51

Jeffries, Dr John 79–80
jet, designing the 120–25, *121*, *123*
jet fighters 126–7

Johnson, Amy 105–7, *106*
kites 76–7
knot speed 58
Knox-Johnston, Robin 71

land speed record 36, 37, 44–6, *45*
land yachts 46
latitude 55
Leonardo da Vinci 78, 140–41, *141*
Levassor, Emile 34
Lincoln, President Abraham 25
Lindbergh, Charles 100, 108–9, *108*
Lindbergh, Erik 108
liquid propane gas (LPG) 160
Liverpool to Manchester railway 21, 22
Lockheed SR-71 Blackbird 133
London and North Eastern Railway (LNER) 27
London Terminal Aerodrome, Croydon, Surrey 97
long-haul flights, birth of (118–49) 98–102, *100*
longitude 55, 56
longships 53–4, *53*
Luftwaffe 118, 122, 127

McAdam, John 12, 14
macadamising 12
MacArthur, Ellen 72
Mach number 132
Maglev (magmetic levitation) train 31
Mary Rose 60, *61*
mathematics 15–16
Mayflower (merchant ship) 65, *65*
measuring speed 139–40
Messerschmitt Me 163B rocket-powered fighter 146, 148–9, *148*
Messerschmitt Me 262 jet fighter 123, *123*, 126, 127

Michelin, Edouard 34
Moller Skycar M400 144–5
Montgolfier brothers 79
motorbikes 49–51, *50*

navigation 55–7, *55*, 112
New Zealand, flights to 117
Newcomen, Thomas 18, *18*
Newton, Isaac 16
Noonan, Fred 110, 111
North American X-15 aircraft 131–2

octane rating 162
ornithopter 78
Ostercamp, Theo 88–9

Pacific route 112–14, *113*
PAL-V gyrocopter 144
Pan American Airways 102, 109, 111, 113, 114, 117
'papyriform' boats 53
parafoil 145
Pearl Harbor attack (1941) 113
Peyron, Bruno 72
Pilgrim Fathers 65–6
pitot head 140
pneumatic tyres 34
power boats 73–4, *73*, **73**, **74**
Power Jets 120, 121, 127
Promontory Point, Utah 25–6, *26*
pulsejet engine 146, 147

Qantas 116, 117

railways
   the American railroads 24–6, *26*
   Channel Tunnel 28–9
   in England 19–23, *22*, **23**
   gathering steam 26–8, **27**
   high-speed trains 29–31, *30*, **30**, **31**, 149–50, 158–9
Rainhill Trials (1829) 21–2

ramjet 150
Reitsch, Hanna 142
relative speeds **140**
Richthofen, Manfred Freiherr von 89
roads
 macadamising 12
 Roman 11–12
Robert Stephenson and Company 20–21
Rocket steam locomotive 22, *22*
rockets 146–50, *147*, *148*
Rolls-Royce 121, 127
Roman roads 11–12
Roper, Sylvester Howard 49
rotating wing machines 140–43, *141*, *143*
Rover company 121
rowing the Atlantic 68–9
Royal Flying Corps (RFC) 87, 88
Rutan, Burt 156

Schneider, Jacques 90
Schneider Trophy races 90–95, *92*
scramjet 150–51, *151*
seaplane (floatplane) 86, 90–91, 94–5, 101
Second World War
 finest piston-engined and first jet fighters **119**
 front-line bombers **119**
 front-line fighters **118**
 heavy bombers **120**
semiconductors 163
sextant 55, *55*
ships
 the Atlantic crossing 64–9
 the Australia route 70–71
 clippers 57–9, *57*
 cruise liners 159
 the fighting ships 59–61, *59*, *61*
 the first steamers 61–3, *63*
 from dug-out to sail 52–7

 round the world sailing races 71–2, *72*
Short-Mayo Composite 101
Sikorsky, Igor 141
Sirius (steam ship) 67, *67*
Skycar 145
slave trade 66–7
solar power 163–4
sound barrier, breaking the 129–30, *129*
South Africa, flights to 103–4
space travel 152–6
 commercial 155–6
 Moon landings 152–3, *152*
 Space Shuttle 154–5, *155*
 visiting Mars 153–4
spark plug 33
speedometer 139–40
stagecoaches 12–15, *14*
steam engine 18–19, *18*, 49, 61, 67
steam power 17–28
 the American railroads 24–6, *26*
 gathering steam 26–8, *27*
 Isambard Kingdom Brunel 23–4, *23*
 railway travel in England 19–23, *22*, *23*
steam turbine 62–3, *63*
Stephenson, George and Robert 22
Stevens, John 25
Sukhoi Su-37 133
supersonic travel 129–32, 136–7

telomeres 154
Terrafugia Transition 144
Tesla Roadster 161
Thames Tunnel 29
tin mining 18
Tour de France 47–8, *48*
Trevithick, Richard 19, 20, 25
Triumph Cycle Company 49
Tupolev Tu-144 ('Concordski') 137

Tupolev Tu-160 Blackjack 133
turbofan engine 125, 127–8, *128*
turbojet engine *125*, 143
Typhoon Eurofighter 132

V-22 Osprey 142–3, *143*
V1 flying bomb (buzz bomb; doodlebug) 146
V2 supersonic missile 146, *147*
Vikings 53, *53*, 64
Virgin Galactic 155–6
Von Braun, Werner 147, 148
Von Ohain, Hans 122, 123

Wallis, Sir Barnes 82
Waterman Aeromobile 144
Watt, James 18–19
wheel, discovery of the 10–11
Whittle, Frank 120–25, 127, 128
wing loading 90, 91
Wright brothers (Wilbur and Orville) 85, 91

X-43A Hypersonic Experimental Vehicle 151, *151*

Yeager, Chuck 130–31, *131*

Zborowski, Count Louis 36, *37*
Zeppelin airships 82
Zheng He, Admiral 54, *54*